CALCULUS

ANALYZING AIRPOWER'S CHANGING ROLE IN JOINT THEATER CAMPAIGNS

Christopher
BOWIE

Fred
FROSTIC

Kevin
LEWIS

John
LUND

David
OCHMANEK

Philip
PROPPER

Prepared for the
United States Air Force

RAND

With the ending of the Cold War, the United States is shifting its military strategy from dealing with the threat posed by the Soviet Union to dealing with the threats posed by regional powers to vital U.S. interests. This strategy will impose new requirements on U.S. military forces. A first step in addressing these issues was the development in 1990/91 of the "Base Force," which proposed reductions in the U.S. military joint force structure and defense budget. The changing world security environment has subsequently produced pressures to reduce U.S. forces and defense spending further. To conduct future force planning, decisionmakers must understand the ability of each component of the nation's military forces—both singly and jointly—to contribute to the new strategy.

This report's main objective is to evaluate the capabilities of future U.S. forces—particularly those force elements provided by the United States Air Force—for achieving key operational objectives in major regional conflicts. Though U.S. national military strategy also contains other important elements, such as deterring nuclear attack and combating terrorism, this report focuses on the capabilities of forces for theater conflicts for one simple reason: As Willy Sutton, the famous bank robber, put it, "That's where the money is." The most costly defense programs and force structure units provide capabilities in this area, and decisions over their future are likely to produce the most intense debates, particularly in light of projected and prospective reductions in U.S. defense spending.

While the analysis in this report focuses on operational concepts and weapon systems that would increase airpower's effectiveness in prosecuting joint-force operations in major regional conflicts, it is clear that a balanced joint military force will be needed to support U.S. national military strategy in the demanding and highly uncertain future environment. Although the analysis does not examine the full range of potential means available to improve the land and sea components of U.S. military power, it also indicates that airpower will become increasingly important in future major regional conflicts because of its rapid strategic mobility and lethality.

Large elements of the combat analysis in this report were conducted using a series of computer spreadsheet models, which provided "transparency" and allowed us to run a very large number of cases for comparative purposes.[1] Many of the factors employed in these models draw upon previous RAND simulation work and analyses of Operation Desert Shield/Storm. Obviously, not all real world conditions can be captured in these (or even more complex) simulations, and this report discusses some of the key limitations.

In addition, an analysis of future forces employs a number of weapon systems currently in development. For most of these systems (such as the C-17, B-2, inertially guided bombs, and the Sensor Fuzed Weapon), the calculations relied upon a body of past and ongoing RAND research into system capabilities, erring as much as possible on the conservative side and employing a parametric approach to bound the areas of uncertainty. Many of these systems, of course, are in the early stages of development, and their actual performance could differ in some ways from the assumptions made for purposes of calculation. Minor variations in performance should not affect the main findings of this analysis.

The research contained in this report was sponsored by the Vice Chief of Staff, Headquarters, United States Air Force and was conducted under the auspices of Project AIR FORCE's Strategy and Doctrine Program. The report's findings should be of interest to all those concerned with future force planning issues.

---

[1]More complex and detailed simulations were employed in certain areas, such as air-to-air combat and attack of surface forces by aircraft.

CONTENTS

FIGURES

TABLES

The collapse of the Soviet Union has transformed the nature of America's "strategic problem." This provides both an opportunity and a need to reexamine U.S. military forces with an eye toward designing a posture suited to the nation's needs in the post–Cold War period. The new emphasis in U.S. military planning on regional conflicts is an appropriate starting point that presents several challenges, including numerous potential locales, smaller forward deployments, short warning times, long deployment distances, and increasingly lethal weapons in the hands of adversaries.

This report's main objective is to evaluate the capabilities of U.S. forces for achieving key operational objectives in future major regional conflicts. In particular, it focuses on means of improving airpower's potential capabilities in the context of joint operations. We recognize that the capabilities of ground- and sea-based forces can be enhanced by new operational concepts and technologies and have highlighted some relevant areas. But we did not explore the full range of options for improving army and naval forces in the same depth.

In conducting this analysis, we first examined the broad outlines of future U.S. national military strategy and a range of potential military threats to U.S. interests. We then developed a number of scenarios involving a major regional conflict between U.S./allied and enemy forces. Employing these scenarios, we simulated the deployment and employment of a joint U.S./allied force to estimate the length of time required to achieve key operational objectives. This approach provides a quantitative and operationally realistic means for comparing various operational strategies and modernization programs in terms of their impact on the U.S. ability to achieve these objectives.

NATIONAL MILITARY STRATEGY

The point of departure for this analysis is the assumption that the United States—a power with global economic interests and interdependencies—must remain engaged in world affairs to influence decision-makers (friend, foe, and nonaligned) as they contemplate choices that impinge upon our national goals. The collapse of the Soviet Union has fundamentally altered U.S. strategy and force planning. But it clearly has not eliminated the need for powerful U.S. forces. In the future, as in the past, the United States must maintain the ability to bring military power to bear when appropriate to protect its interests, as well as those of its allies.

The Joint Chiefs of Staff (JCS) have recommended that the United States should field forces capable of defeating aggression in two concurrent major regional conflicts (that is, conflicts that erupt sequentially, but at times must be prosecuted simultaneously). For purposes here, we took the two-conflict requirement as a given element of our national military strategy and assessed U.S. military capabilities to support the strategy. But such a requirement has important implications for the size and mix of the future U.S. force structure. We would like to make the following observations.

Whether or not one believes the probability is very high of the United States prosecuting two concurrent major regional conflicts, sizing forces for more modest criteria (e.g., for one major regional conflict or for smaller scale conflicts) could engender substantial and unnecessary risks. In the event of a major conflict in one region, such a posture would risk creating opportunities for an aggressor in another region. It would leave the United States vulnerable if a larger threat arose, such as from a coalition of regional powers or the reemergence of an aggressive and anti-Western regime in Moscow. A larger force structure provides flexibility and some margin for responding to the unexpected—both valuable qualities when dealing with something as inherently uncertain as military operations 10 to 20 years in the future. And finally, sizing forces for two conflicts need not cost twice as much as sizing for one, since many key elements of the military infrastructure do not need to be expanded commensurately. But a two-conflict capability will cost more—and hence this requirement is bound to be the subject of much debate in the coming years.

In addition to the gross *quantitative* criterion of being able to prevail in two concurrent major regional conflicts, important *qualitative* criteria should be specified for future U.S. military forces. Economic and manpower constraints, as well as political sensitivities, will constrain the United States from stationing sizable forces overseas on a routine basis. Thus, forces needed to cope with fast-developing crises or conflicts must be rapidly deployable.

When engaged, the United States must be able to achieve its aims in regional conflicts quickly, decisively, and with the capability to minimize casualties. These aims must be clear to three key "audiences": potential aggressors, U.S. allies, and the American public. If the United States lacked such qualitative advantages, the proclivity of other powers to contemplate aggression would be increased, the confidence among its allies that the United States could provide assistance on time undermined, and an American president's options in dealing with future threats to U.S. interests severely constrained.

SCENARIOS

The size of the potential military threat facing the United States has decreased dramatically when compared to the days of the Cold War. The United States no longer needs to plan forces to engage a power with 55,000 tanks, a large "blue-water" navy, and 7,500 combat aircraft. For force planning purposes, analysis of current and potential regional powers indicates that hostile forces confronted in such conflicts could comprise approximately 3,000 to 5,000 tanks, an equivalent number of armored personnel carriers (APCs), between 500 and 1,000 combat aircraft, and perhaps ballistic missiles. Several nations today possess such forces, and other powers possess the economic, military, and technical wherewithal needed to build up to these levels fairly rapidly.

For analytic purposes, we examined a wide range of scenarios involving conflict in Southwest Asia (SWA) and Korea. Neither these nor other scenarios should be seen as predictive of a future conflict, but they are useful as representative future challenges to test the capabilities and robustness of U.S. forces. We varied the warning conditions, environment (in terms of weather and terrain), the size of the opponent's military forces, and the size and modernization levels of U.S. forces to examine sensitivities to outcomes. Given limited U.S. peacetime forward presence in Southwest Asia and the weakness of allied forces there, this scenario proved to be the more demanding— and hence forms the focus for the results summarized below.

Drawing from the analysis of worldwide military capabilities, we postulated an invasion of Saudi Arabia by a force equipped with just over 4,000 tanks and 4,000 APCs (enough to equip 10 mechanized/ armored divisions and an equivalent number of infantry divisions), 500 to 1,000 combat aircraft, and possibly ballistic missiles. Such a force would be expected to quickly overwhelm indigenous allied forces.

The joint force commander's objectives in such a scenario would be to

- Rapidly deploy forces to establish a lodgement;

- Gain local and then theater air superiority to protect arriving forces and establish the conditions needed to conduct effective operations;

- Stop the invading force to minimize loss of territory and vital facilities;

- Conduct strategic strikes to degrade enemy war-fighting capabilities;

- Launch an air-land offensive to evict the aggressor from captured territory.

ASSESSING THEATER FORCES

Warning Assumptions and Force Commitments

For the base case, we focused on the most demanding case: a conflict where "strategic warning" is very limited. Contending with such scenarios raises many challenges. Forces engaged in forward-presence missions play an important role in U.S. military strategy by signaling U.S. resolve and may well deter an adversary contemplating aggression. But if deterrence fails, we believe that for force planning purposes it is both prudent and validated by history to analyze short-warning scenarios. In the past, the United States has often failed to anticipate when and where it has had to go to war. Pearl Harbor in 1941, Korea in 1950, and Kuwait in 1990 were not anomalies—similar failures in gauging the intentions of potential aggressors and responding to strategic warning are likely to be the rule, not the exception, as the United States enters a new era of uncertainty and instability. Moreover, if U.S. forces could deal with short warning scenarios, our analysis indicates they could also deal with conflicts where more warning time is available.

We examined the capabilities of a joint force composed of a contingency corps (5 divisions), 3 to 4 carrier battle groups, 2 Marine brigades (plus attached air), 6 to 10 fighter wings, 80 bombers, and related command, control, and surveillance assets.

Deployment

In our base case, we assumed that no U.S. forces, except for a single carrier battle group, were deployed in theater at the start of the conflict. Accordingly, additional forces would have to be deployed quickly and in quantity. As forward presence declines and conflicts erupt far from U.S. shores, the nation's mobility triad—airlift, sealift, and prepositioning—emerges as an increasingly vital element of U.S. force structure. For the first critical weeks of combat, U.S. forces would have to rely almost exclusively on airlift and maritime prepositioning (the latter greatly increasing U.S. flexibility compared to land-based prepositioning). Sealift remains critical for deployment of heavy forces and for long-term sustainment of all deployed forces.

Establishing Control of the Air

Gaining air superiority would be a top priority for the joint force commander. Control of the air is achieved through establishing a robust air defense network, suppression of enemy air defenses (SEAD), and destroying enemy airfields and command and control facilities.

Given the rapid mobility of U.S. air-to-air fighter forces and their command, control, communications, and intelligence (C³I) support (Airborne Warning and Control System [AWACS] aircraft), we estimate that a robust air defense against aircraft could be established within roughly one week from the decision to deploy. Though others have matched the capabilities of U.S. fighters, the fielding of the AIM-120 medium range air-to-air missile preserves a U.S. advantage in air combat. But active radar missiles similar to the AIM-120 are proliferating throughout the world. Our detailed simulation work indicates that the United States needs to procure a new aircraft to maintain its decisive edge in this critical area. In the air-to-air arena, the United States cannot rely on a missile alone to keep a qualitative advantage.

Ballistic missiles, particularly when coupled with weapons of mass destruction, would complicate operations despite whether these weapons are used by U.S. adversaries. We did not simulate use of such weapons in our analysis. We did, however, allocate sufficient airlift and set deployment priorities to ensure the earliest possible arrival of theater missile defense batteries.

SEAD operations are an integral part of achieving air superiority. To account for this, our simulation allocated a substantial portion (about 25 percent) of the fighter force to SEAD operations. Such an emphasis should enable the United States to minimize attrition of friendly penetrating aircraft. Strikes aimed at destroying airfields and air defense command and control facilities were included in the strategic offensive operations discussed below.

Conducting Strategic Offensive Operations

Conducting strategic strikes against an adversary would degrade an adversary's war-fighting capabilities. Current U.S. capability rests largely upon fighters and sea-launched cruise missiles. The former could be heavily engaged in more pressing tasks, such as stopping a ground invasion; the latter offer a useful, but limited, punch. Equipping the long-range bomber force with precision munitions (such as inertially guided weapons) and standoff weapons (such as cruise missiles) would allow the United States to dramatically increase both the effectiveness of attacks on strategic assets in the early days of conflict and the rate at which it can destroy such targets. Penetrating fighters would still be needed to deal with a range of targets that require great precision and/or a man in the loop, such as hardened facilities and areas where the United States needs to minimize collateral damage.

Stopping Enemy Surface Forces

Stopping enemy surface forces and establishing an "assured defense" (that is, inflicting sufficient attrition on enemy ground forces so that there is a high probability enemy forces would have to stop their advance) depend critically upon the speed at which invading enemy surface forces can be destroyed and disrupted. The analysis examined the contribution of indigenous ground forces, carrier aircraft, and land-based fighters and bombers. In a short warning scenario, land-based airpower would provide the lion's share of this capability. Improving the U.S. ability to stop an invading force depends heavily on fielding dispensers equipped with smart anti-armor submunitions (such as the Sensor Fuzed Weapon or SFW). Our analysis indicates that airpower forces equipped with such weapons could stop a force of 10 armored/mechanized divisions in about one week after the decision to deploy—roughly half the time of the same forces armed with current weapons. Increasing the proportion of airlift assets to deploy fighter forces can further increase the speed at which an assured defense can be achieved. Moreover, inertially guided dispensers filled with smart anti-armor submunitions could be employed by the B-2 bomber to increase up-front punch and further decrease the time needed to stop an armored invasion.

Carrier battle groups provide a unique military presence in peacetime and in crisis may be on scene at the start of conflict. In such cases, their early contribution can be very valuable. But the limited numbers of fighters provided by carriers mean that they can only play a limited role in theater warfare. We examined a range of cases in which carrier fighters were the only attack assets employed—for analytic purposes, we assumed the availability of the USAF's C^3I system to focus carrier firepower most effectively, the use of SFW to maximize kill rates, and typical Southwest Asia weather. In the base case (with three carriers arriving on C+0, C+7, and C+28, respectively), it would take over a month to establish an assured defense.

We also examined two alternatives: one in which four carriers were available and arrived at weekly intervals (C+0, C+7, C+14, and C+21); and one in which four carriers were on station at the start of conflict. In the first alternative, an assured defense could be established in just under four weeks; in the second alternative, the four carriers could establish an assured defense in just over two weeks. Like air and land forces, naval forces cannot be expected to win a war in isolation.

Launching a Ground Offensive

In the case in which an aggressor chooses not to withdraw from captured territory, our analysis indicates that the key constraint in launching a ground offensive is not the ability to weaken enemy forces through attrition, but rather, the rate at which U.S. and allied ground combat and support forces in theater could build up to launch a ground offensive.

After enemy forces had been stopped and dug in, airpower forces could employ current types of point weapons (Maverick and laser-guided bombs) to destroy most of the remaining enemy forces. Our estimates of the time required for this task in the base case range from an additional 8 days (with ten fighter wings in typical Southwest Asia weather) to 27 days (with six fighter wings in typical Korean weather).

U.S. ground forces would require 60 days or more after the decision to deploy to mount an air-land offensive.

Assessing Capabilities for a Second Conflict

Under currently planned force levels (the Base Force), each of the Services would, in principle, possess sufficient residual combat forces for a second conflict (which for analytic purposes was assumed to be of the same size as the first conflict). Of these residual forces, the Army (as currently structured) would have difficulty generating sufficient combat and combat services support to conduct operations in a timely manner for other than light force operations. For short notice operations, the three carriers provided by the Navy are all that typically might be readily available. The Marines could provide an active brigade. The Air Force would retain 10 to 12 wings that might be employed for a second operation, though additional long-range attack aircraft and command and control assets would be needed to conduct the most effective operations.

Most other nations that the United States might be called upon to help defend could provide some sort of ground forces, but most would have greater difficulty in fielding effective offensive air forces. Land-based airpower also appears to offer the most stopping power per commitment of airlift resources. Accordingly, we emphasized deployment of land-based airpower and light U.S. ground forces in our simulation of a second conflict.

The amount of time separating the two conflicts is critical for determining the feasibility of mounting a successful defense. In evaluating U.S. capabilities to deal with two contingencies with D-days separated by *less* than three weeks, our analysis indicates that the strains on the tanker and airlift forces alone would prevent the United States from deploying forces to the second conflict in a timely manner. Conflicts separated by *more* than three weeks would allow it to support operations in the first conflict relying primarily on sealift and shift the bulk of the airlift fleet to deploy forces to the second conflict.

Assuming the latter conditions, our analysis indicates that the United States has the capability in this economy of force operation to blunt an invasion successfully and conduct strategic strikes in a second conflict. The time required to build up additional ground and air forces to eject enemy forces from friendly territory would depend importantly upon the outcome of operations in the first conflict and the availability of sealift assets to close forces to the second conflict.

CONCLUSIONS

Figure S.1 illustrates the contributions over time of the various elements of the U.S. joint force posture. In the early stages of crisis, naval forces provide enduring presence. As we transition to conflict, the relative (but not absolute) contribution of naval forces declines; rapidly deployable land-based airpower emerges as the dominant element in the crucial initial stages of conflict. Ground forces build up slowly but are essential for evicting the aggressor from occupied territory.

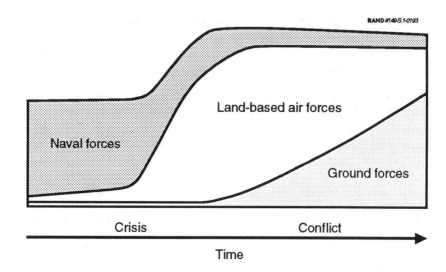

RAND #149-S.1-0193

**Figure S.1—Contribution of Joint Force Components in
Regional Conflicts**

In posturing its forces to deal with short notice theater conflicts, the
United States must rely heavily upon airpower in the crucial initial
stages of combat. Aircraft are highly responsive and mobile, capable
with tanker and airlift support of deploying anywhere in the world in
a matter of days. Such air forces can be supported, at least in the
crucial initial stages of combat, by airlift and can outrange almost
any opponent through use of the nation's tanker fleet. Though attrition cannot be ignored, judicious employment of electronic and
lethal defense suppression systems can minimize losses. Moreover,
air operations place at risk a much smaller number of U.S. personnel
than large-scale ground operations.

These results do not imply that airpower alone will suffice to meet
the needs of U.S. national security. As illustrated by this analysis, in
some situations, weather, terrain, countermeasures, disruptions of
the deployment of forces, and enemy operational strategies could
reduce the effectiveness of an "air dominant" approach. Other scenarios are certainly possible—and such scenarios would stress different elements of the U.S. joint force structure. An insurgency, for example, would typically demand different sorts of forces: advisory
and training missions, civil engineering teams, light ground combat

units, helicopter and fixed-wing gunships, and Special Operations Forces. These results imply that the nation needs a joint land, sea, and air force for use in theater conflicts, which can present potential enemies with the decisive and flexible force needed to underwrite deterrence.

But the results of our analysis do indicate that the calculus has changed and airpower's ability to contribute to the joint battle has increased. Not only can modern airpower arrive quickly where needed, it has become far more lethal in conventional operations. Equipped with advanced munitions either in service or about to become operational and directed by modern C^3I systems, airpower has the potential to destroy enemy ground forces either on the move or in defensive positions at a high rate while concurrently destroying vital elements of the enemy's war-fighting infrastructure. In short, the mobility, lethality, and survivability of airpower makes it well suited to the needs of rapidly developing regional conflicts. These factors taken together have changed—and will continue to change—the ways in which Americans think about military power and its application.

To exploit airpower's potential, the United States needs to ensure its ability to control the air, which allows it to conduct more effective attacks of enemy forces and strategic assets. It needs to equip its future forces with advanced munitions, which play a critical role in enhancing their lethality. Our analysis indicates that procurement of additional long-range fighters capable of carrying heavy payloads (the F-15E) would significantly increase force effectiveness and flexibility. Finally, a rapidly deployable theater C^3I system (consisting of airborne command, control, and surveillance assets combined with deployable ground-based facilities) is essential to the effective operations of these forces—and appears achievable through the integration of current systems if upgraded as planned.

Changes in the international environment combined with the increasing effectiveness of U.S. forces mean that reductions in the U.S. military force structure are both possible and prudent. Future U.S. military strategy will set demanding requirements for U.S. military forces. While a smaller force can support U.S. strategy, that force must be of high quality. Hence, the United States must maintain a "qualitative edge" in its military capabilities through selective modernization.   The enhancements discussed above—mobility forces, advanced munitions, advanced fighters, and C^3I assets—will require a significant investment. It may be necessary to "trade" a portion of the U.S. joint force structure for selective modernization. This will require a new approach to coping with spending cuts, which in the past have focused primarily on reducing procurement accounts and have tended to be apportioned more or less evenly across services and mission areas.

ACKNOWLEDGMENTS

This report was the product of an interdisciplinary RAND study team and the able assistance of a range of Air Force officers on the Air Staff and the Major Commands. Not all who contributed can be mentioned here, but on the Air Staff, we would like to thank General Michael Carns, the Vice Chief of Staff, who sponsored the study. Our thanks also go to Lieutenant General Michael Nelson, the Deputy Chief of Staff for Plans and Operations, and his successor, Lieutenant General Buster Glosson, and Major General Howell Estes III, the Director of Plans, and his successor, Major General John G. Lorber.

Colonel Michael Carlin, Major Scott Dorff, and Major Mark Matthews, who functioned as action officers on the Air Staff during the course of the study, provided valuable intellectual input and assisted greatly in day-to-day administration and coordination. We would also like to pay special thanks to Colonel Charles Miller of the Strategic Planning Division, whose probing questions helped greatly in enhancing the study's analytic richness.

We also profited from the wisdom and expertise of the various Major Commands. At the former Strategic Air Command, we would like to thank Brigadier General James Richards and Theodore Hardebeck; at Space Command, Major General William Jones, Colonel Ray Barker, Lieutenant Colonel James Hale, Lieutenant Colonel James Shoeck, and Lieutenant Colonel Michael Wolfert; at Tactical Air Command (now Air Combat Command), Brigadier General Tony Tolin, Colonel Philip Comstock, Colonel Bob Pastusek, and Lieutenant Colonel David Deptula; and at Military Airlift Command (now Air Mobility

Command), Colonel Jay Marcott, Lieutenant Colonel Alan Jewell, and Michael Ledden.

At RAND, we'd like to thank Brent Bradley and Scott Harris for their careful and insightful technical reviews of an earlier draft. Leland Joe and Dan Gonzales provided key analysis on command and control issues, which could not be presented in this unclassified document. John Green's insights into munitions effectiveness were also critical to the study. Pamela Thompson and Sarah Young performed yeoman work in the text processing of this complex document. Finally, we'd like to thank RAND's superb graphics and publications departments, who put together the complex figures and typeset the final publication.

ABCCC	Airborne Command Control and Communications
ACAS	Airlift Cycle Assessment System
AGM-88	High-Speed Anti-Radiation Missile (HARM)
AIM	Air-intercept missile
AIM-120	Advanced medium-range air-intercept missile (autonomous radar guidance)
AIM-7	Sparrow radar-guided medium-range air-intercept missile
AIM-9	Sidewinder infrared-guided short-range air-intercept missile
ALCM-86C	Air-Launched Cruise Missile 86 (conventional)
AMC	Air Mobility Command
AMRAAM	Advanced Medium-Range Air-to-Air Missile (AIM-120)
ANG	Air National Guard
APC	Armored personnel carrier
ARC	Air Reserve Component
AWACS	Airborne Warning and Control System
C^3I	Command, control, communications, and intelligence
CAP	Combat air patrol
CBU-87	Cluster Bomb Unit 87 (combined effects munition)
CBU-97/B	Cluster Bomb Unit 97/B (sensor fuzed weapon)
CEP	Circular error probable
CONUS	Continental United States
CRAF	Civil Reserve Air Fleet
DMSP	Defense Meteorological Satellite Program
DoD	Department of Defense
DSCS	Defense Satellite Communications System
DSP	Defense Support Program

FEWS	Follow-on Early Warning System
FLOT	Forward line of troops
FY	Fiscal year
GBU-10	Guided Bomb Unit 10 (2000-pound laser-guided weapon)
GBU-12	Guided Bomb Unit 12 (500-pound laser-guided weapon)
GNP	Gross national product
GPS	Global Positioning System
HARM	High-Speed Anti-Radiation Missile
IG/SFW	Inertially guided sensor-fuzed weapon dispenser
IGW	Inertially guided weapon
JCS	Joint Chiefs of Staff
JMEM	Joint Munitions Effectiveness Manual
JOPES	Joint Operations Planning and Execution System
JSTARS	Joint Surveillance, Tracking, and Reconnaissance System
LGB	Laser-guided bomb
MEB	Marine Expeditionary Brigade
Mk 82	Mark 82 500-pound general-purpose bomb
Mk 84	Mark 84 2,000-pound general-purpose bomb
MRC	Major regional conflict
NCAA	Nonnuclear Consumables Annual Analysis
NFIP	National Foreign Intelligence Program
ODS	Operation Desert Storm
PAA	Primary Authorized Aircraft
POL	Petroleum, oil, and lubricants
RSAS	RAND Strategy Assessment System
SAM	Surface-to-air missile
SATCOM	Satellite communications
SEAD	Suppression of enemy air defenses
SFW	Sensor Fuzed Weapon (CBU-97/B)
SIOP	Single Integrated Operational Plan
SLEP	Service Life Extension Program
SSM	Surface-to-surface missile
SWA	Southwest Asia
TLAM	Tomahawk Land Attack Missile
TMD	Tactical munitions dispenser
TOA	Total obligational authority
TSSAM	Tri-Service Standoff Attack Missile

TW/AA	Tactical warning and attack assessment
USA	United States Army
USAF	United States Air Force
USMC	United States Marine Corps
USN	United States Navy
WAS	Wide-Area Multispectral Surveillance Program

Chapter One

INTRODUCTION

The collapse of the Soviet Union has transformed the nature of America's strategic problem. This provides both an opportunity and a need to reexamine U.S. military forces with an eye toward designing a posture suited to the nation's needs in the post–Cold War period. The new emphasis in U.S. military planning on regional conflicts is an appropriate starting point that presents several challenges, including numerous potential locales, smaller forward deployments, short warning times, long deployment distances, and increasingly lethal weapons in the hands of adversaries.

This report's main objective is to evaluate the capability of U.S. forces—particularly those force elements provided by the U.S. Air Force (USAF)—for achieving key operational objectives in future major regional conflicts. In conducting this analysis, we first examined the broad outlines of future U.S. national military strategy and a range of potential military threats to U.S. interests. We then developed a number of scenarios involving a major regional conflict between U.S./allied and enemy forces. Employing these scenarios, we simulated the deployment and employment of a joint U.S. force operating in combination with allied forces to estimate the length of time required to achieve key operational objectives. This approach provides a quantitative and operationally realistic means for comparing various operational strategies and modernization programs in terms of their impact on the U.S. ability to achieve these objectives.

Though U.S. national military strategy also contains other important elements, such as deterring nuclear attack and combating terrorism, we focused on force capabilities for theater conflicts for one simple reason: As Willy Sutton, the famous bank robber, put it, "That's where the money is." The most costly defense programs and force structure units provide capabilities in this area, and decisions over their future are likely to produce the most intense debates.

This analysis was conducted for the Vice Chief of Staff, USAF, who sought insights on ways to evaluate future Air Force capabilities in light of future needs. As a result, the analysis focuses primarily on Air Force capabilities in the context of joint force operations. Though land-based airpower is a critical element of American military power, it is important to remember that it cannot be considered in isolation. U.S. military strategy is joint strategy, and the contribution of any component can only be appropriately judged in that context. Each of the Services provides a set of unique and, in many cases, interdependent capabilities, which we have attempted to account for in this analysis. The capabilities of ground and naval forces can be enhanced by new operational concepts and technologies, and we have highlighted some relevent areas. But we did not explore the full range of options for these forces in the same depth as we did for air forces.

It is a truism that U.S. national security strategy and defense planning are in a period of transition. The United States is, in fact, entering an era around which lie vast uncertainties. Under such conditions, the challenge to the defense planner is to identify the broad contours of the long-term security landscape and adjust accordingly while avoiding, above all, irreversible errors. The analysis that follows is offered in that spirit.

Chapter Two

THE CHANGING STRATEGIC ENVIRONMENT

This chapter reviews the major factors that affect force planning. The first section suggests the broad outlines of future U.S. national military strategy; the second section provides an overview of potential threats to U.S. security. The final section reviews planned changes in the U.S. joint force posture.

U.S. NATIONAL MILITARY STRATEGY

The point of departure of this analysis is the assumption that the United States will remain heavily engaged in world affairs. The collapse of the Soviet Union has fundamentally altered U.S. strategy and force planning. But it clearly has not eliminated the need for powerful and decisive U.S. military capabilities. As a power with global economic interests and interdependencies, the United States must remain engaged in world affairs to influence decision-makers— friend, foe, and nonaligned—as they contemplate choices that impinge upon our national goals. In the future, as in the past, the United States must maintain the ability to bring military power to bear when appropriate to protect its interests, as well as those of its allies.

The United States will continue to rely upon military power to underwrite the following important objectives.[1] Each of these objectives begets an operational strategy:

---

[1]For an overview, see General Colin Powell, *The National Military Strategy of the United States*, USGPO, 1992.

- Deter massive nuclear attack;

- Deter/prevent small attacks with weapons of mass destruction through threat of retaliation, conventional counterforce, and active defenses;

- Deter/defeat regional aggression;

  — Be able to win two "concurrent" major regional conflicts ("concurrent" here means conflicts that erupt sequentially, but must be prosecuted at times simultaneously);

  — Participate in multinational collective security arrangements;

  — Provide security assistance and regional training;

  — Restrict proliferation of threatening weapons and technology;

- Advise and assist friendly governments in countering insurgency, subversion, and lawlessness;

- Deter/prevent terrorist activities;

- Thwart the illegal shipment of drugs into the United States;

- Provide humanitarian relief.

The remainder of this report will focus on only one of these objectives and strategies: defeating regional aggression. We focus on this component for several reasons. The bulk of U.S. training, force planning, doctrinal, and support activities in the future will, properly, center on this component. This category also contains the most costly Department of Defense (DoD) programs and program elements (e.g., A-X, B-2, C-17, F-22, Seawolf), which must be assessed according to their contributions to theater war-fighting capabilities.

None of this is meant to imply that other components of military strategy, such as counterinsurgency and counterterrorism, are lesser included cases of major theater conflict or that forces procured for the latter will necessarily be appropriate for the former. Clearly, the United States will need some specialized assets and force elements to support these components of its national military strategy. But we have not attempted to assess existing or programmed forces in terms of their capabilities for performing these other roles.

Quantitative Criteria

The Joint Chiefs of Staff (JCS) have recommended that the United States should field forces capable of defeating aggression in two concurrent, geographically separated major regional conflicts (MRCs). For purposes here, we took the two-MRC requirement as a given element of our national military strategy and assessed the capabilities of U.S. forces to support the strategy. But such a requirement has important implications for the size and mix of future U.S. force structure. We would like to make the following observations.

Whether or not one believes that the probability is very high of the United States prosecuting two concurrent major regional conflicts, sizing forces for more modest criteria (e.g., for one major regional or smaller scale conflicts) could engender substantial and unnecessary risks. A two-MRC requirement would help avoid the risk of opening a "window of opportunity" to an aggressor in one region during a period when U.S. forces are committed to a major conflict elsewhere. Sizing for two contingencies would also provide an important hedge in the event that a larger threat (e.g., a hostile Russia, China, or a coalition of smaller states) arises, or in the event that an MRC or other situation evolves in an unpredictable way requiring U.S. operational improvisation. And a U.S. force structure capable of dealing with two MRCs could play an important role in deterrence, since it would also be large enough to help deter potential adversaries from seeking to build forces with capabilities equal to (or greater than) the United States. Such would not be the case for a U.S. force posture capable of handling only one MRC.

The U.S. ability to forecast future force needs has been far from perfect: Peak U.S. force deployments in Korea, Vietnam, and Iraq exceeded planners' prewar expectations by a factor of two in critical areas. The sheer size of U.S. forces gave it needed flexibility in these conflicts, since each was different in some important ways from the wars for which the United States had planned. The composition of the forces needed also differed from the balance of the total force. For example, though the United States only used 30 percent of USAF fighter assets in Operation Desert Shield/Storm (ODS), some elements of the force (e.g., long-range fighter bombers, elements of the C^3I structure) were almost totally committed. In the new era, a conflict like Desert Storm could require the commitment of 50 percent or more of the USAF. If U.S. preconflict planning were accurate, this would allow it an adequate reserve. But if the nature of the war were different from that anticipated, forces in critical areas might be more heavily committed and the margin of error substantially reduced.

Finally, sizing for two MRCs need not cost twice as much as sizing for one. Most proposals examining the future military force posture focus on numbers of "shooters": various types of ground force divisions, such naval vessels as carriers, surface combatants, submarines, and fighter and bomber wings. Actually, the direct cost of building, manning, and maintaining these assets tends to be a fairly small element of the overall cost of the force. The training infrastructure, research and development facilities, and other important elements do not rise proportionately with increases in force size above a rather low level. Likewise, some capabilities, such as airlift and sealift, aerial refueling, space-based and theater-based command and control assets, and the like, are needed almost regardless of force size. Indeed, many of the assets needed to contend with a single conflict are in fact costly capabilities with forcewide utility and either do not require duplication for a second MRC or require only modest increases in selected areas. But a two-MRC capability will cost more—and hence this requirement is bound to be the subject of much debate in the coming years.

Qualitative Criteria

In addition to these gross quantitative considerations, future strategy suggests several important qualitative desiderata. For example, rapid deployability will become increasingly important in the future. Economic, political, and manpower constraints, combined with domestic political pressures and foreign sensitivities, will constrain the United States from stationing sizable forces overseas on a routine basis. In the past, the United States has often failed to anticipate when and where it has had to go to war. Pearl Harbor in 1941, Korea in 1950, and Kuwait in 1990 were not anomalies—similar failures in gauging the intentions of potential aggressors and responding to strategic warning are likely to be the rule, not the exception, as the

United States enters this new era of uncertainty and instability. This heightens the importance of speed of response.

Forces that are ready and capable of conducting decisive operations quickly at great distances from the United States are essential to the credibility of its national strategy. Without them, U.S. national security strategy would not be credibly supported and deterrence would be weakened. The capability of these forces must be clear to three key "audiences": potential aggressors, allies, and the American public. The United States must not only be able to achieve its aims in regional conflicts but must be able, with high confidence, to win quickly, decisively, and with the capability to minimize casualties. If the United States lacked such qualitative advantages, the proclivity of other powers to contemplate aggression would be increased, the confidence among its allies that the United States could provide assistance on time undermined, and an American president's options in dealing with future threats to U.S. interests severely constrained.

WORLDWIDE MILITARY CAPABILITIES

The former Soviet Union still possesses large numbers of weapons, and Russia, at least, has the material wherewithal to threaten the United States and its allies. The analysis that follows is predicated upon our assessment that for the next decade or more, Russia and the other states of the former Soviet Union will be heavily preoccupied with domestic economic crises and political turmoil. As a result, it is highly unlikely that these states would constitute a serious external military threat or see it as in their interest to attack their neighbors beyond the borders of the former Soviet Union.

Nonetheless, it is still far from clear that the various democratic experiments now underway will succeed. Should an authoritarian leadership regain power in Moscow, it might attempt to reestablish Russia's internal empire. To accommodate these possibilities, sizing U.S. forces for two regional conflicts can provide a hedge against the reemergence of a large-scale Russian military threat. This also implies that the United States should strive to draw down its forces gradually over the next decade or so to provide the maneuvering room needed should events in the former Soviet Union dictate that more attention be paid to it.

Military forces possessed by potential regional powers are far smaller in size than those of the former Soviet Union but are far from trivial. Moreover, the relatively small scale and uneven professionalism of most Third World militaries create conditions that can lead to the frequent emergence of serious imbalances in regional military alignments. Under such conditions, when crises or conflicts arise, the security of friendly nations can deteriorate quite rapidly.

In Operation Desert Shield/Storm, the United States and its allies deployed very sizable forces to defeat a well-equipped regional aggressor. Forces the size of Iraq's can present a serious challenge, particularly if they are determined and well led. The United States cannot reduce its forces on the assumption that future foes will be inept. Additionally, the potential proliferation of weapons of mass destruction (nuclear, biological, and chemical) adds an ominous dimension to the equation. Complicating the problem is the fact that the United States must typically deploy forces far from its shores when going to war—and must be able to do so on short warning. In conducting theater force analysis in future contingencies, the United States cannot simply add up its overall force structure and compare it to those possessed by regional powers. Instead, it must assess the capabilities of its forces to deploy to the fight and conduct operations over time in the face of various threats.

Figure 1 provides a perspective on the changes in the relative magnitude of the military threats the United States might have to face in the future. The United States and other allied forces are shown to provide reference points. The breakup of the USSR and the Warsaw Pact means that U.S. force planning no longer needs to focus primarily on contending with a power possessing 55,000 tanks and 7,500 combat aircraft. Several nations today can field 3,000 to 5,000 tanks and an equivalent number of armored personnel carriers (APCs). And many nations possess the military, economic, and technical wherewithal to build up to these levels fairly rapidly, particularly with the amount of military hardware on the market in the world today.[2] In terms of force structure, a nation with just over 4,000 tanks and 4,000 APCs could field, for example, about ten ar-

---

[2]For example, Iran is currently in the process of a major arms buildup.

mor/mechanized divisions complemented by an equivalent number of infantry divisions.[3]

Figure 1 illustrates that many nations today can field 500 to 1,000 combat aircraft. Typically, half of these can conduct surface attack missions—the remainder, air defense operations. Few nations possess true multirole aircraft because of the costs of equipment, training, and required munitions inventories.

By comparing the number of tanks to aircraft, Figure 1 also illustrates that the United States, relative to most other countries, has invested quite heavily in airpower.[4] Nations with modest economic resources and the political will and motivation can field sizable ground forces, which are manpower intensive, useful for internal security and nation building, and typically less costly compared to other forms of military power. Through conscription and investment, land forces can grow in size over a relatively short period. Air forces are more difficult to construct due to the sheer cost of equipment and the sorts of technical skills and training needed to make them effective fighting forces. To effectively employ air forces in offensive operations requires developing and maintaining a costly C^3I infrastructure—something few nations currently (or are likely to) possess. As a result, few countries will have the potential to match the United States in terms of airpower, particularly offensive airpower.

The demise of the Soviet Union has also changed the global naval balance significantly. As shown in Figure 2, after Russia, whose fleet is rapidly losing operational readiness, the U.S. Navy today possesses more major warships than the next two largest navies present today combined. And of the ten largest navies after the United States, seven belong to U.S. allies. When considering displacement, quality, and firepower of forces, the balance is even more lopsided. The difficulties nations encounter in developing effective air forces are heightened when attempting to develop naval forces that can contest

---

[3]To relate these numbers back to U.S. force posture requirements, it would take several months to deploy sufficient U.S. ground forces to stop an advancing army of this size by themselves.

[4]Combat aircraft counted include Air Force, Navy, and Marine fixed-wing assets. The U.S. inventory data shown here include large numbers of tanks held in storage.

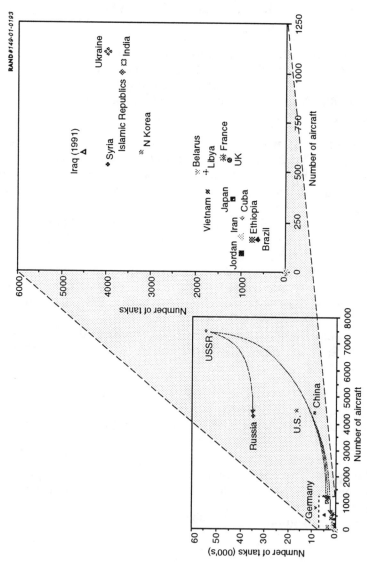

SOURCE: *The Military Balance*, London, International Institute for Strategic Studies, 1991.

Figure 1—Worldwide Military Capabilities Remain Significant

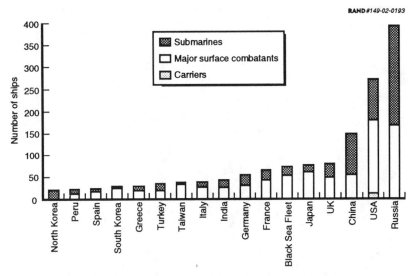

SOURCE: The Military Balance, London, International Institute for Strategic Studies, 1991.

Figure 2—The World's Navies, 1991

control of the seas. As a result, the newly emerging naval balance, even after the United States reduces its naval posture to planned levels, is unlikely to change for decades because of the time required to develop, train, and field "blue water" naval forces. Most of the naval threats the nation might face over the next two decades will likely come from small coastal navies or land-based airpower operating with anti-ship weapons.

Figure 3 shows developing nations that are pursuing improved ballistic missile capabilities. One of the lessons of the Gulf War is that Third World militaries can employ such weapons in an operationally effective manner despite the best efforts of the U.S. military. The strategic and operational significance of ballistic missiles will increase dramatically if these missiles are coupled with warheads of mass destruction. A number of nations—Iraq, Iran, and North Korea, for example—are pursuing both ballistic missiles and weapons of mass destruction. Technologies for improving the capabilities of ballistic missiles are spreading, and coping with these

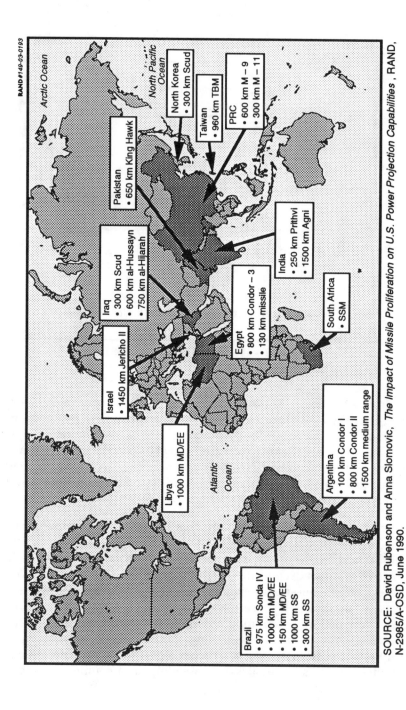

Figure 3—Missile Development Programs

SOURCE: David Rubenson and Anna Slomovic, *The Impact of Missile Proliferation on U.S. Power Projection Capabilities*, RAND, N-2985/A-OSD, June 1990.

weapons will be a key challenge for U.S. forces in the future. Contending with a theater ballistic missile threat is a reality the United States must be prepared to deal with in the new world.

LOOKING TO THE FUTURE

As the Soviet threat began to collapse, the Bush administration took the first steps to reduce the size of the U.S. military in late 1989, when it undertook a closely held review of minimum long-term force requirements for the future. President Bush introduced, in general terms, the new concept—the "Base Force"—at an address in Aspen, Colorado, in August 1990. The next day, Iraq invaded Kuwait, and the President's speech and its implications were largely overlooked. Subsequently, in early 1991, the Chairman of the Joint Chiefs of Staff, General Colin L. Powell, outlined what he considered to be minimum military force posture objectives for the emerging security environment.

Table 1 shows, for purposes of comparison, the overall U.S. force posture at the end of FY91, alongside estimates of the planned posture at the end of FY97. Under the Base Force, plans call for reducing Army total division equivalents by 35 percent; Navy combatants (carriers, surface warships, and submarines) by 13 percent; Marine brigade equivalents by 27 percent; and Air Force fighter wings by 29 percent. Overall personnel levels would decline 20 percent. This smaller force would require a smaller proportion of the nation's resources: Recent Department of Defense testimony states that the Base Force is estimated to require a commitment of 3.4 percent of the nation's gross domestic product in FY97 (resulting in a budget of about $235 billion in $FY92) compared to FY91's 5.6 percent ($282 billion in $FY92).[5]

But many issues still remain. The Base Force was developed when the Soviet threat still remained and when U.S. strategy for the post–Cold War world was in transition. With the devolution of that threat, many of the old principles for force planning clearly need to be re-

---

[5]See Statement of the Secretary of Defense, Dick Cheney, before the Senate Armed Services Committee, January 31, 1992.

Table 1

U.S. Military Forces

Service	Force Element	End FY91	End FY97 (est)
Army	Active division Equivalents[a]	17.6	12.6
	Reserve division Equivalents[a]	17.3	10
Navy/Marine Corps	Carriers (total/deployable)[b]	14/12	12/12
	Surface combatants (active/reserve)[c]	153/34	144/16
	SSNs	87	78
	Active fighter wing Equivalents[d]	14.7	13.3
	Reserve fighter wing Equivalents[d]	3.1	2.8
	USMC brigade Equivalents[e]	11	8
Air Force	Active fighter wing Equivalents[f]	24.9	15.7
	Reserve fighter wing Equivalents[f]	12.9	11.3
	Bombers	261	184

[a]Includes separate brigades and regiments.

[b]Under new Navy policy, all carriers, even those undergoing extended modernizations, are counted as deployable. The training carrier, AVT-59, maintains a limited mobilization capability as well.

[c]Reserve total includes frigates in mobilization reserve.

[d]In terms of 72 Primary Authorized Aircraft (PAA). Includes KA-6D, EA-6B.

[e]Based on combined active and reserve infantry battalions.

[f]In terms of 72 PAA. Includes EF-111s and RF-4s. Does not include CONUS air sovereignty assets.

examined. Indeed, as the architect of the Base Force, General Powell, has testified: "We purposely designed the Base Force to be able to adapt and adjust to a rapidly changing world. Obviously, as that world changes, our strategy and accompanying force structure will change with it."[6]

In addition, there is continuing pressure to reduce spending on defense below levels specified by the Bush administration. The key question, however, is whether the United States can posture the nation's force structure to support U.S. national military strategy at an affordable cost. To address that question requires a deeper understanding of the contributions of the various elements of the nation's joint force structure to achieve key objectives in major regional conflicts.

---

[6]Statement of General Colin Powell, Chairman of the Joint Chiefs of Staff, before the Committee on Armed Services, United States Senate, January 31, 1992.

Chapter Three

THEATER FORCE EFFECTIVENESS

In this chapter, we

- Describe the conditions that we believe the United States should plan for in future regional conflicts;

- Outline our approach toward analyzing U.S. military capabilities;

- Assess capabilities to deploy forces to a distant theater of operations;

- Estimate the ability of U.S. airpower (both land- and sea-based) to achieve key operational objectives in the critical early weeks of a conflict;

- Incorporate airpower capabilities into an evaluation of joint force capabilities in achieving theater campaign objectives;

- Assess capabilities to execute a second MRC;

- Identify high leverage force enhancements.

For purposes of analysis, our study started with elements of the planned 1997 Base Force. We examined potential conflicts in the years 1997 to 2010 to assess the contributions of various future capabilities and systems now in development. To analyze these forces, we started with two different scenarios for our force effectiveness analysis: a conflict in Southwest Asia (SWA) and a conflict in Korea. Neither these nor other scenarios should be seen as predictive of future conflict, but they are useful as representative future challenges to test the capabilities and robustness of U.S. forces. We varied the warning conditions, the environment, the size of the opponent's military forces, and the size and modernization levels of U.S. forces to examine sensitivities in outcomes. In this summary of our work, we present highlights of our scenario analysis for a confrontation in the Gulf region because this proved to be more demanding for U.S. forces than was a conflict in Korea.[1]

Figure 4 depicts the conditions under which U.S. forces might have to operate. It represents the study's basic scenario for U.S. force planning and evaluation. As noted previously, a number of countries today can field 3,000 to 5,000 tanks, a similar number of APCs, and 500 to 1,000 aircraft, and many nations possess the infrastructure needed to rapidly build up to these levels. A nation with just over 4,000 tanks, 4,000 APCs, and an accompanying force of artillery could field an adversary force of the size illustrated here: 10 armored/

RAND #149-04-0193

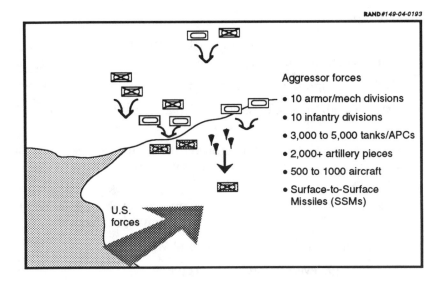

Aggressor forces

• 10 armor/mech divisions

• 10 infantry divisions

• 3,000 to 5,000 tanks/APCs

• 2,000+ artillery pieces

• 500 to 1000 aircraft

• Surface-to-Surface Missiles (SSMs)

U.S. forces

Figure 4—Postulated MRC Threat, Late 1990s

---

[1]While the Korean scenario remains a relevant and useful case for force planners, it is far less demanding than the SWA scenarios employed in our analysis. The reasons for this include the powerful and rapidly mobilizable forces of the U.S. ally; the presence of sizable, prepositioned stocks of equipment and supplies for deploying U.S. forces; and the highly developed in-place C^3I structure.

mechanized divisions, and an equivalent (or greater) number of light divisions.[2]

In the postulated scenario, an aggressor launches short-notice air and ground attacks against an outnumbered neighboring state. U.S. forces, not present in the region at the outbreak of the attack, would need to deploy, supplement indigenous forces, establish a foothold in the theater, and then stop and defeat the aggressor.

The first tasks for U.S. forces in such a situation would be to secure ports, bases, and lines of communication needed to reinforce and operate in the theater. In operational terms, the joint force commander's highest initial priorities would be to protect friendly forces from air and ballistic missile attack while concurrently blunting the enemy's land invasion—i.e., to stabilize the situation while securing a lodgement in the region. And from the outset, U.S. forces would seek to attack strategic assets in the hostile state's homeland to isolate the leadership from forces in the field, degrade the aggressor's war-fighting capabilities, and reduce its will and ability to continue aggression. The order of priorities naturally could change in different situations—the forces selected for deployment and employment should have sufficient flexibility to respond in strength to changing conditions.

Next, U.S. forces would set out to create the conditions needed to achieve victory. These would include gaining theater air superiority, continuing to destroy enemy surface forces and the supplies that sustain them, and increasing the tempo of strategic attacks against the aggressor's war-making capability.

Finally, the United States must be prepared to impose a satisfactory conclusion by forcing the enemy from the occupied territories when necessary with a coordinated ground and air offensive, while taking measures to limit the adversary's future war-fighting potential.

Other scenarios are certainly possible—and such scenarios would emphasize different elements of the U.S. joint force structure. An insurgency, for example, would typically demand different sorts of forces: advisory and training missions, civil engineering teams, light ground combat units, helicopter and fixed-wing gunships, and

---

[2]In the Persian Gulf, the postulated adversary might be Iran or Iraq. In Korea, it would be the Democratic Peoples Republic of Korea.

Special Operations Forces. In other scenarios, detection of threatening activity combined with appropriate political decisions might provide sufficient time to mount a power projection demonstration and/or a joint deployment of land, sea, and air forces to deter conflict. Such cases might resemble Operation Desert Shield.

This analysis, however, focuses on perhaps the most demanding scenario: a conflict where "strategic warning" (or more precisely, the ability or willingness to act on ambiguous warning indicators) is very limited. This case is highly demanding, but we believe that for force planning purposes it is both prudent and validated by history. Moreover, if U.S. forces can deal with this situation, our analysis indicates that U.S. military forces could also deal with scenarios where more warning time was available.

FORCE COMMITMENTS

In conducting our analysis, we initially laid out joint service force commitments to the theater. For our base case, we assumed that no U.S. forces (other than one carrier battle group) were deployed in the region prior to the conflict (i.e., C-Day, the day on which the decision is made to deploy and commit forces, is equal to D-Day, the day on which combat begins). Given this modest forward presence, the time required to deploy U.S. forces is a critical element in the assessment of their combat effectiveness. Our analysis factored in the contributions of the joint "mobility triad": airlift, sealift, and prepositioning (both sea-based and land-based).

Airlift forces are the fastest mobility asset but are limited in terms of the cargo that can be carried in terms of weight, size, and volume. In our base case, we assumed employment of 90 percent of the organic airlift fleet (10 percent of the force being withheld for other missions) plus the second stage of the Civil Reserve Air Fleet (CRAF II) starting at C+0.[3] This would require some preparations prior to C-Day, such as a reserve call-up. We also ran cases where the availability of the airlift fleet increased more gradually (as in Operation Desert Shield)

---

[3]As of August 1992, CRAF I consists of 18 long-range international passenger aircraft and 22 long-range cargo jets. CRAF II consists of 77 long-range international passenger aircraft and 39 long-range cargo jets. CRAF III consists of 252 long-range international passenger aircraft and 141 long-range cargo jets. CRAF Stage III was not used in our base cases.

and where enroute and destination base constraints were encountered (see below).

Sealift provides very high volume but is much slower than airlift. In our analysis, we assumed ships were ready to begin loading operations on C-Day. Again, this would require some preparations prior to the commencement of hostilities. Assuming sailings out of the continental United States (CONUS), approximately three weeks would elapse before the first vessels arrived in theater. The assumption that vessels are ready to begin loading on C-Day is a best case assumption—if not, the United States would be more dependent for a greater length of time upon airlift and prepositioning.

The limited volume of airlift and the slow speed of sealift heighten the importance of prepositioning. We assumed the availability of maritime prepositioning ships based within nine days sailing time of the theater. These vessels would contain not only ground force equipment, but Air Force munitions and selected equipment as well. In terms of land-based prepositioning, we assumed that no heavy ground force equipment was positioned on land in the theater. For the Air Force, until the maritime prepositioning vessels arrived and their cargoes were distributed, all "preferred" Air Force weapons (AIM-9 short-range air-to-air missiles, AIM-120 medium-range air-to-air missiles, modern anti-armor weapons, and precision guidance kits for bomb bodies) were airlifted into theater.[4] Only general purpose bomb bodies (Mk 82s and Mk 84s) were assumed available in the theater at the outbreak of conflict (though we also ran cases where no land-based prepositioning whatsoever was available).[5] Carrier task forces were assumed to be resupplied from accompanying logistics vessels.

The analysis examined the deployment and employment of the following major combat units, which would be available with the Base

---

[4]Off-line analysis indicates that about 100 C-130 airlifters (or, in the future, a mixture of C-17 direct delivery airlifters and C-130 airlifters) could distribute USAF munitions to bases throughout theater should no trucks be available for intratheater movement.

[5]Requiring airlift of all "preferred" weapons is a fairly conservative assumption. For example, in July of 1991, the United States sold 2,000 Mk 84s, 2,100 CBU-87s, 770 AIM-7s, and laser guidance kits to Saudi Arabia. As in Desert Shield, it is likely the Saudis would make these available to U.S. forces. See "New World Orders: U.S. Arms Transfers to the Middle East," *Arms Control Today*, March 1990, pp. 34–35.

Force. We varied the size and, in some cases, deployment rates of these forces to probe for key sensitivities in the analysis.

- **Army:** A U.S. Army contingency corps of five divisions: the 82nd Airborne, the 101st Air Assault Division, the 24th Mechanized Division, the 1st Cavalry Division (armored), and the 7th Light Infantry Division.[6] The closure schedule for these units is discussed in more detail below. In addition to the combat component of U.S. Army forces, combat and combat service support at echelon above division had to be deployed to sustain combat operations. Considering an "immature" theater infrastructure, a minimum of 180,000 combat support/combat service support personnel (along with their equipment) would have to be mobilized and deployed to sustain the deployed corps.[7]

- **Navy:** Three carrier battle groups (each of which contains a carrier, two cruisers, four destroyers, two attack submarines, and related components of an underway replenishment group).[8] We assumed each carrier was outfitted with 20 air-to-air F-14As, 40 F/A-18C multirole fighters, and 5 EA-6B jamming aircraft.[9] The battle group's associated submarines and surface combatants carried typical loadouts of Tomahawk Land Attack Missiles (TLAMs) for attacks against some classes of land targets. One carrier task force was assumed to be on station at C+0, the other two arrived on C+7 and C+28, respectively. We also ran variations examining the sensitivity of results to carrier availability. In

---

[6]See *Base Force—The Required Force to Execute the National Military Strategy*, by Captain Britten, Headquarters, United States Army, June 1992. As noted in this paper, "Analysis indicates that the minimum force necessary to establish the building blocks for the initial force packages is five fully structured divisions." A stronger ground force mix would include another heavy division instead of a light infantry division. Because this is the last element of the contingency corps to arrive in theater—and would close in the second wave of sealift—this change would not affect our results or conclusions.

[7]This requirement is based upon analysis of support requirements for Operation Desert Shield/Storm.

[8]The Base Force concept envisions 12 deployable carriers, which would yield three available in short-notice contingencies. See *Seapower for a Superpower*, Headquarters, United States Navy, 1992. Typically, two-thirds of an underway replenishment group are required for each forward deployed battle group. As outlined in official Navy posture statements, four carriers are typically required for each carrier forward deployed (though the number actually available depends on distances from homeports).

[9]Other aircraft on the carriers—S-3s, E-2Cs, and helicopters—take up the rest of the available deck and hanger space.

one case, we assumed four carriers arriving at weekly intervals (C+0, C+7, C+14, and C+21).[10] In another, we assumed four carriers on station at the start of conflict.

- **Marines:**   Two Marine brigades and their associated air components (each containing 24 F/A-18Cs, 40 AV-8B Harriers, and 6 EA-6B jamming aircraft). The first was assumed to close at C+11 (using equipment from the maritime prepositioning ships), the second at C+21 from the CONUS. Because the Marines have maritime prepositioned equipment, they are a likely choice for early deployment to provide needed ground-fighting power.

- **Air Force:**   Ten fighter wings (each numbering 72 Primary Authorized Aircraft or PAA),[11] 80 heavy bombers (16 PAA B-2s and 64 PAA B-1Bs), tankers, tactical airlift, and an array of command and control assets.[12] The fighter wings (FWs) in our base case consisted of 1.6 FWs of air-to-air F-15Cs for air superiority missions; 1.3 FWs of F-15Es, 0.5 FWs of F-117As, and 0.8 FWs of F-111Fs for long-range attack operations; 5.3 FWs of multirole F-16Cs; and 0.5 FWs of EF-111 electronic jamming aircraft.[13] Portions of the F-15E and F-16C force were employed for defense suppression operations.   We also ran cases examining the commitment of fewer FWs (six and eight, respectively) and a range of bomber options.

U.S. forces were assumed to operate in conjunction with indigenous forces (Republic of Korea ground and air forces in the case of Korea, Saudi and Kuwaiti ground forces in the case of Southwest Asia), but no allied forces from outside the region were committed to the bat-

---

[10]Four carriers is the maximum number that is likely to be available at the Base Force levels for an MRC with less than one month of warning.

[11]Historically, the USAF has deployed an average of ten fighter wings to the three major post–World War II conflicts: Korea, Vietnam, and Iraq.

[12]These include such assets as E-3 Airborne Warning and Control System (AWACS), the E-8 Joint Surveillance, Tracking, and Reconnaissance System (JSTARS), RC-135 Rivet Joint electronic surveillance aircraft, U-2 reconnaissance aircraft, EC-130 Airborne Command Control and Communications (ABCCC), and the ground-based command and control system needed for joint operations.

[13]This force was developed through consultation with experienced operators to provide a balanced mix of force capabilities needed to prosecute future air campaigns. As discussed in subsequent sections, increasing the number of long-range interdiction aircraft (such as F-15Es) proved to increase the effectiveness of land-based fighter operations.

tle. This appears prudent for planning purposes. If additional forces were available, these would serve to improve friendly force capabilities.

Based upon analysis of planned airlift fleet capabilities (see below for more details), we determined that the following force elements could be airlifted to the theater before the first sealift vessels began arriving from the CONUS: aerial port units to support airlift operations, the 82nd Airborne Division, nine Patriot batteries for both air defense and theater ballistic defense, a combat aviation brigade from the 101st Air Assault Division, a C^3I system, and logistics support, personnel, and preferred munitions for the ten fighter wings. These forces were thus delivered using airlift, as were personnel from the Marines, who were flown in to "marry up" with equipment delivered by maritime prepositioning ships.

Sealift ships would begin arriving about three weeks after C-Day, delivering the majority of the remaining forces and most of the consumables needed to sustain the operation of all deployed forces. This highlights the fact that in a short warning scenario, any force employed in the first three weeks or so (the actual time depends on distance from home ports) must be in place, operate from the CONUS, or be deployed and supported by airlift and land-based/sea-based prepositioning.

MODELING TOOLS

Analysis was performed using three linked spreadsheet models: a force deployment model, an airpower force employment model, and a ground combat simulation. Data obtained through RAND research into Operation Desert Shield/Storm helped in determining the consistency and accuracy of our results.

To determine force closure rates, we used a modified version of the Air Mobility Command's Airlift Cycle Assessment System (ACAS) model (which factors in, among other things, types of aircraft; mission capable rates; crew constraints; base constraints; and bulk, oversize, and outsize cargo requirements).[14] We enhanced this

---

[14]Bulk cargo can be carried by all aircraft. Oversize cargo can be carried only by C-141s, wide-body commercial aircraft such as B-747s and DC-10s, and wide-body

model to better account for base constraints and other factors. The contributions of sealift and prepositioning were factored into the analysis as well.

The deployment model provided an estimate of the time when aircraft would arrive in theater. To produce an estimate of combat potential, we developed an airpower force employment model. This model simulates the employment of aircraft in combat using mission capable and sortie rates consistent with those achieved in the Gulf conflict. Munitions effectiveness was calculated using the appropriate weapons effects manuals (with the data adjusted to reflect the real-world experience of Operation Desert Shield/Storm). The effects of command and control were simulated by increasing the probability of an aircraft arriving on target when the required command and control systems had deployed to theater and had begun operations.[15] Environmental conditions, such as weather and terrain, were simulated by changing target acquisition parameters (for example, in poor weather, fewer aircraft would be able to locate targets).

We used a variety of detailed simulations to conduct off-line analysis in key areas. For example, to simulate attacks on enemy surface forces with advanced munitions, we employed the Army's highly detailed JANUS model. Similarly, to explore the effects of changing equipment and concepts on air-to-air engagements, we employed the Air Force's high fidelity TAC BRAWLER air combat simulation.

We employed the airlift and airpower force effectiveness models to analyze airpower performance in the pursuit of three operational objectives:

- Air superiority;

- Destruction of strategic assets;

- Destruction of enemy surface forces (chiefly armored units).

---

dedicated airlifters such as C-5s and C-17s. Outsize cargo can only be carried in C-5s and C-17s.

[15]We also relied upon operational experience and detailed simulations conducted in other RAND studies to inform our assumptions about the contributions of surveillance assets and C^3 to overall force effectiveness. These effects were captured in assumptions regarding deployment requirements for key assets (such as E-3B/Cs, E-8s, RC-135s, and ABCCC, among others), engagement rates, the degradation of air defenses, and the probability of target acquisition.

The output of this model regarding the third of these objectives was then incorporated into a detailed, situational evaluation of specific air/land campaigns to determine the point at which a successful defense was likely to be achieved. This methodology, developed by RAND, provides a series of step-by-step assessments of the outcome of air/land campaigns considering terrain, the relative combat power of ground units, movement rates, and the contributions of airpower. In essence, it is a more sophisticated version of tabletop war games and employs computers to assist in tracking units and their combat power, as well as assessing the outcome of engagements. We have compared the results of this situational evaluation with RAND's Strategy Assessment System—a global war game utilized by many agencies of the U.S. defense community—for a set of controlled cases. The results were consistent.

Finally, in every scenario, we tracked the number—and, through use of the combat models, the capabilities—of all U.S. combat units not employed in the MRC that would be available for operations elsewhere. Figure 5 illustrates some of the key variables and assumptions employed in the analysis. The shading in the boxes represents the relative level of effort (i.e., the heavier the shading, the greater the depth of analysis). In terms of scenario assumptions, we varied the mobilization time, the time frame (to incorporate the contributions of future systems and force modernization), and the size and nature of the threat. In regard to the latter, for example, we examined Southwest Asia and Korea scenarios, an SWA scenario with double the threat ground forces, and an SWA scenario with Korean weather. We also examined the implications of varying the time span between the onset of one conflict and the start of the second MRC.

In each of the cases examined, we also explored the impact of a range of assumptions regarding logistics support, the types and numbers of "shooters" (i.e., bombers and Air Force, Navy, and Marine fighters), differing munitions, and airlift force levels. Overall, about 350 separate cases were run.

FACTORS NOT EXPLICITLY CONSIDERED

The objective of our effort was to determine joint force effectiveness in representative scenarios to assist in future force structure development. Because of the uncertainties involved in preparing for future conflicts (such as threat, location, and time) and the rather gross

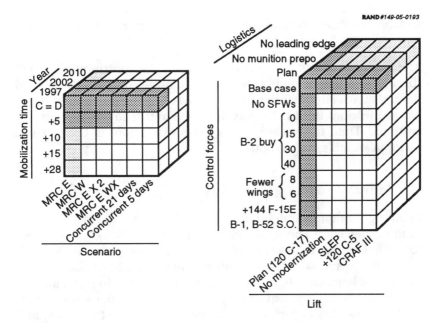

NOTE: "Plan" refers to the assumed availability of general-purpose bomb bodies in theater. "Leading edge" refers to an Air Force deployment concept where sufficient support is brought in to support a unit for one week of operations; follow-on support is brought in before those supplies are exhausted.

Figure 5—Project Analysis Plan

level of the calculations, the analysis did not formally take account of several factors that could have important effects on a "real world" operation. These included the factors discussed in the subsections below.

Detailed Contributions of C³I Systems

Capable C³I systems offer very high leverage, but current analytic tools are inadequate to quantify satisfactorily the effects of changes in C³I support. We factored in the deployment requirements of appropriate numbers of the E-3 AWACS, the E-8 JSTARS, RC-135 Rivet Joint electronic surveillance aircraft, U-2 reconnaissance aircraft, EC-130 ABCCC, and the complete ground-based system required to support theater operations in our deployment calculations. The deployed system would provide command and control capabilities for air, ground, and naval forces engaged in theater combat.

The analysis reflects best judgments about the contributions of C^3I through the use of multipliers. For example, the effectiveness of ground attack aircraft is degraded (by reducing the probability of target acquisition) until JSTARS is deployed and ready to support operations. Thus, the analysis treats the contributions of C^3I assets as an input variable, not an output.

U.S. Military Planning Capability

Planning for rapidly unfolding contingency operations is a complex, multifaceted task. Though specific plans exist for likely scenarios, we found in Operation Desert Shield/Storm, for example, that the tactical and operational situation did not match existing plans. In a rapidly evolving crisis/conflict, planners might need to plan and execute massive deployment and employment operations simultaneously. The ability to perform these tasks depends on the quality and flexibility of existing plans, the use of automated planning aids, and the training, knowledge, and availability of planning staffs in the theater and CONUS. Operation Desert Shield/Storm pointed out many strengths and weaknesses of the current joint planning system. We did not conduct a specific analysis of the ability of the U.S. military to plan deployment and employment operations in the future, but we assumed that the joint planning system would function effectively.

Employment of Weapons of Mass Destruction (Nuclear, Biological, and Chemical)

The proliferation of weapons of mass destruction and delivery vehicles would complicate operations whether or not these weapons are used by U.S. adversaries. We did not simulate use of such weapons in our analysis. We did, however, take note of the need to protect arriving U.S. forces along with allies from such threats by allocating sufficient airlift and setting deployment priorities to ensure the earliest possible arrival of air-to-air fighters for the defensive counter-air role, theater missile defense batteries for ballistic missile defense, and related elements of the C^3I structure. Separate studies are under way at RAND on the implications for U.S. national security strategy and military operations of adversaries equipped with weapons of mass destruction.

Petroleum, Oil, and Lubricants (POL) Availability

The availability of POL is unlikely to be a constraining factor for the explicit scenarios (SWA and Korea) considered here because of the substantial military and economic infrastructure that already exists in both of these areas. Indeed, intertheater distribution and storage could be more of a problem than availability. Though developed infrastructures have grown—and will continue to grow—as commercial air and sea traffic expand throughout the world, in other scenarios, we recognize that POL availability could be a limiting factor in the early days of a campaign before sealift (tankers) arrive in the theater of operations. The availability of POL for high-tempo air, land, and sea operations in a major conflict is an area that deserves further study.

Explicit Attrition Analysis

The state of the art in estimating attrition in air or surface operations leaves much to be desired, and this project did not attempt to advance it. Models are able to show useful relative differences for different aircraft and threats, but they have not provided accurate estimates of absolute levels. Moreover, given the inherent uncertainties associated with planning forces for the next 20 years, it seems unrealistic to aspire to providing a serious analysis of attrition. To perform such analysis, we would need a detailed order of battle for enemy threat systems along with an assessment of operator performance for defensive systems—data simply not available given the uncertainties involved. Moreover, even with such input, it is unclear how useful or reliable the results would be. Nonetheless, we recognize that minimizing attrition is a key desideratum. To do so, in all of our simulated campaigns we allocated sizable portions of the available forces to the establishment of air superiority at the beginning of the campaign (air superiority operations require air-to-air forces, defense suppression assets, and attacks against airfields and C^3I facilities). Until that was achieved, the simulations degraded force performance[16] according to the expected status of enemy air and surface defenses.

The Effects of Enemy Aircraft, Missile, and Terrorist Attacks on Arriving Forces, Bases, Airfields, and Ports

An important factor that bears on the employment of airpower in an MRC is the potential for attacks on airbases to disrupt operations. We did not conduct a detailed assessment of these effects but relied on previous analyses.

Because we recognize the importance of these factors, we included, as noted earlier, the establishment of a robust air and missile defense as a first priority to protect arriving and allied forces. Such a defense would first be concentrated around the most critical areas and then would expand as additional aircraft and surface-to-air missile batteries arrived. Most nations the United States might be called upon to defend possess some air defense capabilities, which could provide some initial protection against air attacks. Carrier-based air-to-air fighters would also help build a more extensive air defense in the early stages. And both indigenous and carrier air defenses could be rapidly supplemented by Air Force fighters, which, along with their AWACS support, are highly mobile. Additionally, the elements of the command and control system for the defense of critical assets and areas in the theater were scheduled very early in the postulated deployment flow.

Patriot batteries, or other future missile defense systems, were also among our first priorities for deployment to defend against ballistic missiles as well as to augment defenses against aircraft. With respect to unconventional threats, security forces for the defense of airbases were an inherent part of the fighter wing deployment packages. The latter could be augmented by U.S. and indigenous ground forces.

Vulnerability to airbase attack is often thought to be a primary limitation of USAF and Marine fixed-wing aircraft operations. Extensive modeling analyses at RAND and other institutions over the past decades have illustrated that this vulnerability is probably over-

---

[16]For example, by reducing the number of aircraft that successfully deliver their ordnance on target.

stated. In the wake of the devastating Israeli strike against Arab air forces in the 1967 Arab-Israeli War, hardened facilities, additional takeoff and landing surfaces, and runway repair apparatus have proliferated throughout the world. As a result, operations at modern combat air bases can be degraded, but are extremely difficult to shut off for any length of time—at least when considering conventional weapons.

From a first-order examination of the airfields in SWA and Korea, we assess that a prudent enemy planner would conclude that it would require more than a squadron size raid (18 to 24 aircraft) to successfully arrive at a base and shut down operations for a meaningful period (say, 12 to 24 hours). And to close the runways at an airfield successfully with ballistic missiles would require missiles far more accurate than today's Scuds, along with advanced runway penetrating submunitions. Even with assuming such weapons, a substantial number of Scud-type ballistic missiles would be needed to shut down operations at an airfield for 12 to 24 hours. As was seen in the Gulf War, attempts to shut down airfields can be extremely costly in terms of time and effort—even when conducted by a force possessing air supremacy.

Basing vulnerability can also be alleviated by taking advantage of a unique U.S. national asset: its tanker force. The large U.S. aerial refueling fleet provides its forces with an inherent advantage over most other nations: the ability to outrange an opponent. With air refueling, U.S. air forces can operate from a widely dispersed set of bases and carriers beyond the range of most threat aircraft and ballistic missiles. The increase in range, however, would reduce sortie rates and increase the amount of time required to achieve key theater objectives.

Finally, it is not only airbases that need to be protected. In general, if an enemy has the ability to successfully attack airbases, it would also have the ability to attack ports, marshaling areas, logistics depots, naval vessels, and other critical assets. These vulnerabilities can best be ameliorated through the rapid deployment of air superiority and ballistic missile defense capabilities—a priority emphasized in our deployment plans.

Base Access and Base Availability

A related issue is the availability of suitable bases for all U.S. forces—ground, sea, and air. Of course, ascertaining base availability involves inherent uncertainties when considering combat operations and the state of geopolitics many years in the future. In terms of theater access, if a friendly or allied nation were attacked or threatened, it is highly likely that the nation would grant access to U.S. forces (as was seen in the Gulf, a region where base access had previously been considered highly questionable). In addition, a large number of airbases capable of supporting combat operations exists around the world, and these would provide a hedge against the risk of being denied access to the immediate locale of conflict. As distance from the area of interest on the earth's surface increases, so does the number of bases and governments with which to negotiate base access agreements.

Aircraft range plays a critical role in alleviating the uncertainties associated with base access. The greater the combat range of an aircraft, the more likely it is to find a suitable beddown base in any given theater. Tanker aircraft, of course, also play a critical role in extending the range of all USAF, United States Navy (USN), and United States Marine Corps (USMC) aircraft. In short, the greater the combat radius of U.S. forces, the larger the number of bases that are potentially available, the larger the number of governments available to negotiate base access arrangements, and the less vulnerable U.S. forces are to attacks by shorter-ranged opponents. But it must again be remembered that increases in range would decrease the effectiveness of U.S. forces.

DEPLOYING FORCES TO THE THEATER

The first step of our assessment of U.S. theater forces was to "deploy" them to the theater. We chose as a base case to assume that the only U.S. combat force present in the theater at the outbreak of conflict was a single carrier battle group; C-Day, the commencement of deployment, is equal to D-Day, the commencement of combat operations. We also examined as excursions cases in which the United States began reinforcing the theater some days or weeks prior to the attack in response to strategic warning.

Figure 6 shows a sample output of our deployment model for forces and mobility assets available in 1997. We employed 90 percent of the organic U.S. airlift fleet currently planned to be available at that time (in terms of Primary Authorized Aircraft [PAA], 110 C-5s, 198 C-141s, 36 C-17s) and the civil aircraft that could be readily mobilized in the second stage of the Civilian Reserve Air Fleet agreement (CRAF II consists of 51 long-range cargo and 95 long-range passenger aircraft). The remaining 10 percent of the military airlift force was withheld for other national missions. On average, our simulation resulted in approximately 75 military airlift sorties per day landing in theater (compared to Operation Desert Shield experience of about 70 sorties per day).[17] In accordance with Operation Desert Shield experience, we allocated 5 percent of the available airlift to deploy support for airlift operations, 50 percent to the Army, 20 percent to the Navy and Marines, and the remaining 25 percent to the Air Force.[18] These track closely with Operation Desert Shield allocations, as seen in Table 2.

Such an airlift would require call-up of reserve component airlift and tanker crews and aerial port squadrons. We also ran a variation in which the airlift fleet "ramped up" to full capacity over a period of four days (as in Operation Desert Shield); this delayed the closure of some Air Force squadrons, for example, by about two days, compared to the base case. However, reallocating the airlift fleet over the first four days to emphasize the deployment of Air Force assets would restore fighter closure times to those of the base case and only delay the deployment of the first Division Ready Brigade of the 82nd Airborne by one day.[19]

---

[17]The utilization rates we used are quite close to those experienced in Operation Desert Shield. For the C-5, the RAND model used 6.30, while the rate was 6.11 for Operation Desert Shield. For the C-141, the RAND model used 8.84, while the Operation Desert Shield rate was 8.01.

[18]As noted previously, preferred munitions for air forces to sustain combat operations were airlifted until maritime prepositioned munitions arrived (around day nine); general purpose bombs were assumed to be prepositioned in theater in our base case.

[19]The reallocation of the airlift force to deploying Air Force units (not including airlift support) would be as follows: C+0, 88 percent; C+1, 59 percent; C+2, 44 percent; C+3, 35 percent; C+4, 29 percent; C+5, 29 percent; C+6, 29 percent; C+7, 27 percent. This allocation would hold until C+15, when the Air Force share would decline to 25 percent (again, not including airlift support, which would consume an additional 5 percent).

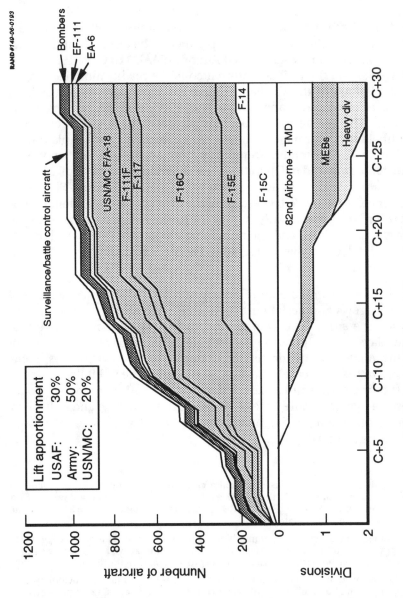

Figure 6—Joint Force Closure, SWA

Table 2

**Allocations of Airlift Force
(percent of committed force)**

Service	RAND Model	Operation Desert Shield
Army	50	46
USN/MC	20	24
USAF	25	25
Airlift force	5	5

Figure 6 above also reflects our baseline assumption that one Navy carrier battle group would be on station at the outset of the conflict, with a second closing on C+7, and a third on C+28. We also examined a case in which one carrier was on station at C+0 and three additional carriers arrived on C+7, C+14, and C+21; and a case in which four carriers were on station at the start of the conflict. The two Marine brigades, along with associated air units, arrived on C+11 and C+21. Marine ground force equipment and munitions were brought in by sealift (using maritime prepositioned and CONUS-based ships); personnel arrived by air.

Airlift Modernization

The United States needs a large organic military airlift fleet to close forces at this rate. Today's fleet, augmented by CRAF Stage II assets,[20] can deliver approximately 3,100 short tons per day to Southwest Asia, assuming the same airfields were available as those employed in Operation Desert Shield/Storm. Early in the next decade, the majority of the C-141s that form the backbone of the Air Mobility Command (AMC) fleet will reach the end of their service lives.[21] If they are not replaced, organic airlift capacity would be reduced by about 50 percent. As shown in Figure 7, the planned acquisition of 120 C-17s (102 PAA) would provide an organic fleet ca-

---

[20]This assumes CRAF participation levels as in 1991. Changes in participation rates, the type of aircraft committed, or the definition of stages would alter these values.

[21]About 30 percent of the C-141 fleet is believed to have operational lifetimes beyond this time period.

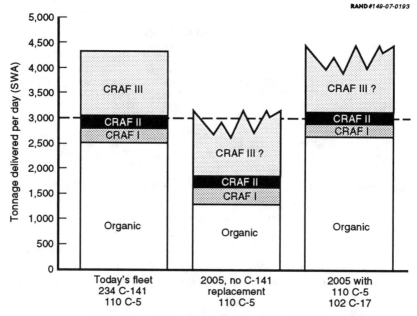

RAND#149-07-0193

Assumes same basing constraints as Operation Desert Shield

Figure 7—Airlift Capabilities

pable of delivering slightly more than today's fleet (though with a greatly increased outsize cargo capability).[22]

The potential future contribution of CRAF III is shown in jagged boxes because of continuing uncertainties over its availability and capacity. In general, the larger and more successful U.S. airlines have become more reluctant to sign up for CRAF for two main reasons: fears that a call-up could severely disrupt their market shares (particularly if competitors were not called up as well) and the increasingly limited peacetime government business available as U.S.

----

[22]Current plans also call for the retention of 60 PAA C-141Bs, which would increase the total tonnage delivery capability by about 275 tons over that shown on the chart (as well as increasing U.S. airdrop capability). But because the C-141 force is reaching the end of its fatigue life, this contribution will be eliminated by the middle of the next decade or so.

forward presence declines.[23] Compounding these problems are the types of commercial aircraft that will be available in the future. Projections of future aircraft orders by U.S. carriers indicate an increasing emphasis on smaller B-767–class aircraft (due to the attractiveness of hub and spoke operations) at the expense of large aircraft such as the B-747, which offer much greater airlift capacity for long-distance operations.

Airlift capacity is highly sensitive to basing constraints. Though much attention tends to focus on the type and number of airlifters employed, the aircraft must be considered as part of a system. Basing constraints can have a major impact on throughput. For example, limits on the number and/or capacity of enroute bases decrease the number of aircraft that can land, refuel, and launch; similar problems can affect theater destination bases. During Operation Desert Shield, for example, over 60 percent of the airlift flow was funneled into a single base (Dhahran) during Phase I of the deployment.[24] This factor, in combination with others, limited arriving sorties to only 70 per day compared to the force's anticipated capability of 120 sorties per day. Enroute and staging base constraints did not prove to be a major problem during Desert Shield, but could be encountered in the future. Accordingly, the United States must consider throughput capability in the face of these constraints.

Several options are available to maintain U.S. airlift capability: extend the life of C-141s through a Service Life Extension Program (SLEP), procure more C-5s, rely on CRAF III, or, as planned, procure C-17s. Figure 8 illustrates the throughput capacity of these options in three cases: (1) the same level of base availability as was employed in Operation Desert Shield; (2) constrained enroute basing (i.e., no access to either Torrejon or Rhein-Main); or (3) constrained base availability in the area of operations (i.e., only Dhahran and Jubail available).

As can be seen in Figure 8, constraining base availability greatly reduces throughput for the SLEPed C-141 option and the C-5 option. In the case of constrained enroute base availability, the CRAF option appears attractive because airliners could theoretically use commer-

---

[23]In return for CRAF commitments, airlines are provided with preferential government travel and business contracts.

[24]August 8 through November 4, 1990.

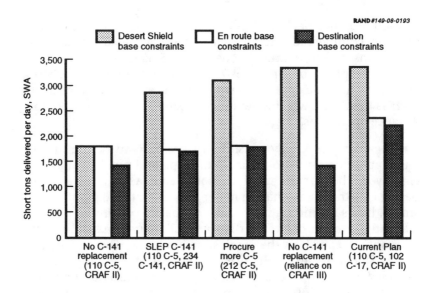

RAND #149-08-0193

Figure 8—Airlift Modernization Options

cial enroute fields. However, sovereign nations can be skittish about allowing large volumes of hazardous material to flow through their commercial airports—in Desert Shield, for example, Germany, Belgium, and the United Kingdom severely limited the number of aircraft carrying munitions and other dangerous cargo that could transit civil fields. Restrictions were also placed in regard to night operations. The CRAF approach would also not solve the problem of basing constraints in the theater of operations. At the same time, it would greatly increase U.S. dependence upon increasingly uncertain access to aircraft of the commercial fleet.

With the planned C-17 option (assuming the aircraft meets performance goals), our simulation indicates that organic capacity increases by an estimated 17 to 36 percent compared to the C-5 or C-141 options when base access is limited. The C-17 consumes about the same amount of ramp space as a C-141, but its wide-body configuration allows it to carry more cargo and its on-the-ground maneuverability allows more aircraft to flow through bases. The capability to sustain throughput in situations where basing is constrained will become increasingly important in a world in which overseas force commitments are reduced and alliance structures may be in flux.

The advantage of the C-17 is its flexibility, i.e., its capability to conduct both long-range strategic missions and shorter-range tactical missions. If greater emphasis is placed on the strategic mission, ongoing RAND analysis suggests that planners may wish to consider a mix of C-17s and the latest variant of the B-747 (the 400 series). The latter aircraft offers advantages in range and payload. However, it is dependent upon large well-developed airfields and specialized loading and unloading equipment.

Finally, airlift capabilities cannot be considered in isolation. Each element of the "mobility triad"—prepositioning, sealift, and airlift—offers unique advantages and disadvantages. To support its national strategy, the United States must attempt to capitalize on each method's virtues to compensate for the others' limitations. In the SWA scenario, nearly all combat power for at least the first three weeks is dependent upon prepositioning and airlift.

Airlift is the most responsive and flexible mobility tool, but its relatively small volume heightens the importance of earnestly pursuing both land-based and maritime prepositioning efforts. Three to four squadrons of maritime prepositioning vessels deployed around the globe and loaded with ground force equipment, logistics support assets, and air-delivered munitions would offer great flexibility and improve force responsiveness and effectiveness in a wide range of scenarios. A complementary responsive theater distribution system is also essential to maximizing prepositioning's contribution to joint force mobility. Also, for sustained operations, sealift—and the protection of sea lines of communication—will remain an essential and critical element of the mobility triad.

FORCE EFFECTIVENESS ANALYSIS

With this background on the deployment analysis in mind, we turn to highlights of our force effectiveness assessments. As noted above, the general scenario postulated involved an invasion of Saudi Arabia by ten mechanized/armored divisions and a similar number of infantry divisions. In such a rapidly unfolding crisis, airpower (both land-based and sea-based) would be the primary force that the United States could deploy and employ quickly enough to blunt the initial attack (though some degree of ground opposition would be useful in channeling the enemy's forces to increase their vulnerability to air attack). Once the enemy assumes a defensive posture (is forced to stop advancing due to attrition and/or disruption of the attack), more time would be available to allow ground forces to deploy to the theater and move into position. In such a conflict, airpower would have three primary tasks:

- Achieving air superiority

- Destroying strategic targets

- Stopping and defeating enemy surface forces.

The allocation of air forces to these tasks would shift over time in response to operational plans and the battlefield situation. Initially, U.S. forces would have to establish a foothold and secure ports, bases, and marshaling areas. Employing a large share of air assets to air defense, defense suppression, and stopping invading ground forces would be high priority missions at the outset. The relative weight of effort devoted to strategic air offensive tasks and other missions would depend on the specific operational and tactical conditions. As the situation in the ground battle area stabilized, an increasing weight of effort could be devoted to strategic offensive missions.

Achieving Air Superiority

As has been consistently highlighted by history since the advent of modern airpower, air superiority—control of the air—provides strategic, operational, and tactical freedom of action while denying these advantages to the opposing side. Without control of the air, all land, sea, and air forces must attempt to operate exposed to air attack, something increasingly difficult to do in the face of modern airpower. Simply put, air superiority is a prerequisite to the effective conduct of joint theater operations and would be a top priority of any joint force commander.

Achieving air superiority is reached through several means

- Establishing a robust air defense network

- Suppression of enemy air defenses

- Destroying enemy airfield facilities and command and control network.

Ideally, these tasks should be accomplished simultaneously, but the phasing and weight of effort devoted to each task is dependent on the resources available and the specific situation. Success in one area abets the effort in the others. For example, reduction of an enemy's air attack potential through strikes against an enemy's airfields and C^3I system reduces required effort in the air defense area. Similarly, defense suppression efforts increase the effectiveness of strikes against airfields (as well as other types of targets).

Figure 9 shows our concept for the establishment of a robust air defense network to protect allied and arriving forces. Indigenous assets, if available, would provide some initial protection, as would fighters from the carrier battle group present at the outset. U.S. Army Patriot batteries could provide ballistic missile and area air defense of critical operational zones. Arriving air-to-air fighters would flesh out the air defense network. Each squadron has the capability of manning autonomous layered combat air patrol (CAP) orbits and of supporting CAPs with flights at various stages of ground alert around the clock. Typically, one squadron can protect an avenue of approach to a critical area. By the time three full air-to-air squadrons are operating with AWACS, these aircraft, supplemented by naval fighters, should be able to provide theaterwide coverage for the MRC scenarios we investigated. Additional aircraft are needed to perform critical force protection missions (i.e., escort and fighter sweeps) and bolster the air defense network if the enemy possesses the capability of mounting massed attacks. Backup from the multirole force provides depth and flexibility to an air defense if required.

Figure 10 shows the buildup of the forces that would constitute an air defense network in an MRC. The time required to establish a robust air defense is dependent upon a number of factors, such as the number of air-to-air capable squadrons deployed, the ability of AWACS to concentrate forces in critical areas, the success of offensive counterair missions in constraining the enemy's ability to conduct massed attacks, and the qualitative superiority of U.S. and allied air-to-air forces.

The buildup of an air defense system is facilitated by the mobility of the elements of the network. Air-to-air fighter squadrons are rapidly deployable, and the weapons needed to arm these aircraft are light

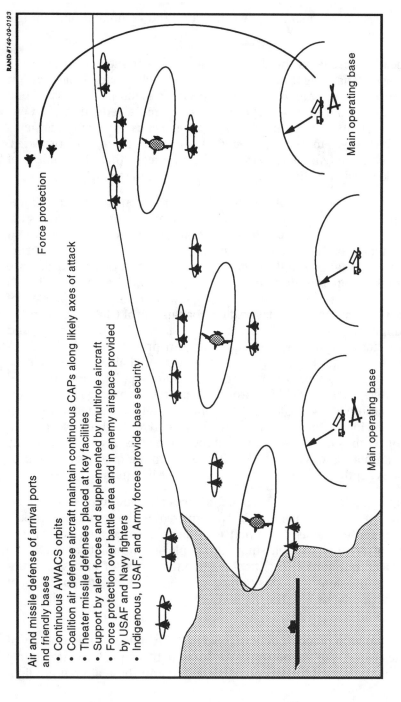

Figure 9—The Air Battle

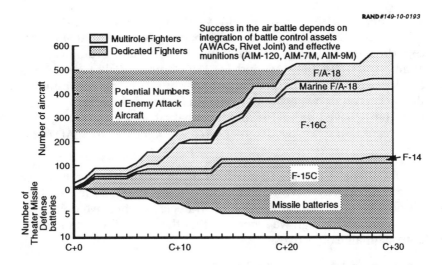

Figure 10—Sufficient Forces for Air Defense and Force Protection Achieved Early, 1997

(and in our simulations, airlifted into theater as necessary). Similarly, AWACS aircraft are rapidly deployable and can provide an autonomous command and control platform for air defense and other air superiority missions (such as offensive fighter sweeps). Patriot batteries (or, in future, a follow-on theater air/ballistic missile defense system) could also be airlifted in quite rapidly. Given these factors for the scenarios and threats considered in this analysis, a robust air defense network appears achievable in less than a week from the start of deployment.

Suppression of enemy air defenses (SEAD) is an integral part of achieving air superiority. Suppressing air defenses through electronic combat, self-protection, and lethal means may become a more demanding role in the future. Regional aggressors may realize that it is difficult to challenge U.S. forces directly in the air as was seen in Operation Desert Storm. However, proliferation of surface-to-air missiles (SAMs) and anti-aircraft artillery may be a lower cost alternative—and one which may prove attractive because of the availability of these systems throughout the world.

Without a specific defense laydown, we could not optimize SEAD forces. We did allocate a significant portion of capable deploying assets to SEAD missions. Specifically, we dedicated up to 24 F-15Es, 48 F-16s,[25] and 12 F/A-18s per available battle group as lethal SEAD assets (that is, delivering anti-radiation missiles and ordnance). In addition, EA-6Bs, EF-111s, and EC-130 Compass Call assets would conduct jamming and spoofing of enemy radar and communications.[26] Overall, we estimated that sufficient defense suppression could be achieved within the first two weeks if, as we simulated, 20 to 25 percent of the deployed fighter force was allocated to SEAD operations.[27]

Finally, offensive counterair missions are an integral part of the quest for air superiority. We have included these missions for analytic purposes under the examination of strategic air offensive objectives in this project, though we recognize that these missions span across both campaign objectives.

Maintaining Air Superiority Over the Long Term. Success in the air is, of course, also very dependent upon a qualitative advantage. The decisive edge is gained and retained by a combination of realistic, demanding training and advanced equipment. The generation of U.S. fighters produced in the 1970s and 1980s, manned by well-trained pilots, has had a qualitative and operational edge over potential adversaries for many years. But other fighters currently available on the market, such as the Su-27, the MiG-29, and the Mirage 2000, essentially match the aerodynamic performance of U.S. first-line air superiority aircraft. Moreover, new fighters now in development (the European Fighter Aircraft, the French Rafale, the Japanese FSX, and other possible new fighters from the former Soviet Union) have the potential to further erode the U.S. edge should these enter the inventories of potential adversaries.

---

[25]Currently, F-15Es and F-16Cs do not have the capability of providing the HARM missile with range information, which greatly increases the missile's lethality. This capability is possessed by the aging F-4Gs, but current Air Force plans envision necessary software and hardware modifications on F-16s and F-15s to provide a ranging solution for the HARM missile. Navy and Marine Corps F/A-18s also do not currently possess ranging capability.

[26]As a point of comparison, in Operation Desert Storm, the USAF deployed 62 F-4G Wild Weasels as its primary lethal SEAD asset—USN and USMC F/A-18s, A-6s, A-7s, and EA-6Bs also contributed at varying levels during the course of operations.

[27]This force allocation for SEAD is consistent with Operation Desert Storm and has been validated by operators on the Air Staff.

But changes in airframe are not the only factors to consider. The air-to-air fighter force employed in our simulations differs in one distinct aspect from the force deployed and employed in Operation Desert Shield/Storm. By 1997, U.S. forces will have sufficient stocks of the AIM-120 Advanced Medium Range Air-to-Air Missile (AMRAAM) to provide a significant qualitative edge over most likely opposing air forces. Detailed combat simulations indicate that equipping the U.S. fighter force with AMRAAM would reestablish U.S. qualitative superiority for some time to come. However, an advantage based upon a missile alone can be short-lived: Published reports indicate that as many as six other nations (Japan, Russia, the UK, India, France, and Taiwan) are developing active radar missiles similar to AMRAAM—and others may gain access to the necessary systems and technologies.[28] Proliferation of active radar missiles would significantly erode the qualitative edge possessed by U.S. forces and, given the importance of early air superiority to joint force operations, create serious risks.

Figure 11 illustrates the progression of the effectiveness of USAF fighters compared to potential adversaries in the air-to-air mission over time. The results shown here illustrate classified simulations run using a highly detailed dynamic computer model in support of this project.[29] Until recently, F-15Cs with AIM-9 Sidewinders and AIM-7 Sparrow missiles would have enjoyed a substantial advantage in terms of exchange ratios. However, enemy fighters equipped with similar missiles would lead to an even fight based on equipment. Arming the F-15s and other U.S. fighters with the AIM-120 missile restores the qualitative edge for the U.S. fighter force in the air. But, if other nations field active radar missiles, the advantage would disappear. Moreover, the air-to-air arena would become much more lethal. Active radar missiles have the potential to revolutionize long-range air combat much as the all-aspect infrared missile did for the close-in fight.

When we incorporate a next-generation platform with capabilities similar to the F-22, the simulations indicate that the U.S. edge in the critical air-to-air mission would be restored. In the course of this study, we have run simulations for the full spectrum of air-to-air

---

[28] *Forcecast International,* DMS Market Intelligence Report, November 1991.

[29] We employed the TAC BRAWLER model, which is widely used in the defense community for these sorts of simulations.

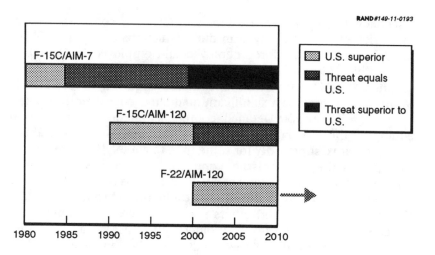

RAND #149-11-0193

Figure 11—Maintaining Air Superiority

combat conditions and weapon system combinations—and the results follow the same general trends.

Potential adversaries' ability to field active missiles and aircraft as
advanced as the current U.S. generation is not in question: The
technology is available and will proliferate. A more debatable issue is
whether the air forces of future potential adversaries will be able to
field aircrews with the sort of advanced skills currently maintained
by U.S. forces. Maintaining current U.S. levels of highly realistic
training is an essential ingredient to a U.S. air-to-air advantage, as is
the continued development and deployment of advanced missiles.
Over the long term, however, the detailed simulations indicate that a
new fighter would be needed to maintain control of the air.

Destroying Strategic Targets

Early destruction of the enemy's leadership, command and control
assets, industrial infrastructure, lines of communication, and other
key war-fighting capabilities can help ensure a decisive victory in
war. Strategic attacks both reduce the enemy's ability to conduct war
and affect its strategic calculus. Attacks against command and control centers and such strategic targets as airfields would also play a
role in the air superiority battle.

The general concept of operations for attacking strategic assets in the enemy's homeland is shown in Figure 12. In our 1997 scenario, portions of the long-range fighter bomber force, TLAMs, and a limited number of standoff weapons fired by USAF long-range bombers (the conventional variant of the air-launched cruise missile [ALCM-C]) would conduct the bulk of these early attacks. Most other ground attack aircraft in the theater would be tasked to stop the advance of enemy surface forces to stabilize the situation (an operation described in the following section). If possible, a larger emphasis on strategic offensive operations is desirable, but it depends heavily on the character of the initial phases of the conflict. As the campaign progresses and the invasion is blunted, the weight of effort dedicated to the strategic offensive campaign would increase: long-range fighters (such as the F-15E and F-117) and heavy bombers would conduct these attacks.

The postulated attack force would combine stealth assets, saturation with cruise missiles and decoys, and manned aircraft capable of defending themselves penetrating at low altitude and protected by defense suppression and air-to-air forces. This combination of different attack assets and penetration profiles would greatly complicate an enemy's defense problem. At present, only the United States could present such an attacking array.

The number and characteristics of strategic assets possessed by potential adversaries will vary over time and from country to country. Preliminary analysis suggests that Iraq is fairly representative of medium-sized Third World nations. In Operation Desert Storm, the United States attacked over 700 strategic targets (which presented a total of about 3,000 aimpoints). Of those requiring precision weapons (~1,000 aimpoints requiring less than ten meter accuracy), about 25 percent were "time critical" (i.e., C^3I nodes, radar sites, leadership facilities, etc.), meaning that there were important operational benefits to be gained by destroying them as soon as possible.

The total number of aimpoints attacked over the course of a campaign could be significantly larger than 3,000 as the enemy repairs damage (thus requiring additional attacks), disperses its assets, and conducts deception operations. We thus analyzed how long it would take for the joint force to attack 3,000 to 5,000 distinct aimpoints, 1,000 of which required precision attack (that is, with weapons whose circular error probable [CEP] is less than 10 meters). In

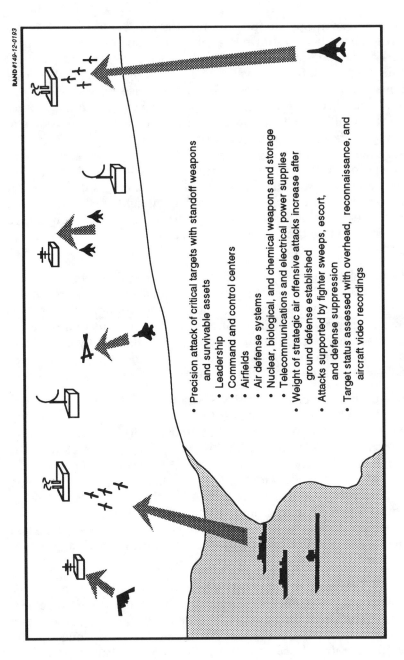

- Precision attack of critical targets with standoff weapons and survivable assets
  - Leadership
  - Command and control centers
  - Airfields
  - Air defense systems
  - Nuclear, biological, and chemical weapons and storage
  - Telecommunications and electrical power supplies
- Weight of strategic air offensive attacks increase after ground defense established
- Attacks supported by fighter sweeps, escort, and defense suppression
- Target status assessed with overhead, reconnaissance, and aircraft video recordings

RAND #149-12-0193

Figure 12—Strategic Air Offensive

different situations, the number and character of targets would certainly vary, but a target array of this size is representative.

Figure 13 shows the cumulative number of aimpoints that could be attacked by programmed U.S. forces in 1997 during the first 20 days of a strategic air offensive campaign. The number of assets available to conduct these attacks is constrained by deployment rates, the need to destroy attacking armored forces, and requirements for defense suppression. Only fighter and TLAM capabilities are illustrated in this figure.[30]

The rate at which strategic aimpoints could be struck depends heavily upon the allocation of attack resources (there might be situations where strategic strikes would be more critical than dealing with enemy surface forces). In our simulations, however, at the outset of the campaign we allocated only 20 percent of the available long-range air-to-surface fighters (e.g., F-15Es) and all F-117s (36 PAA) to strategic air offensive operations. After a successful ground defense

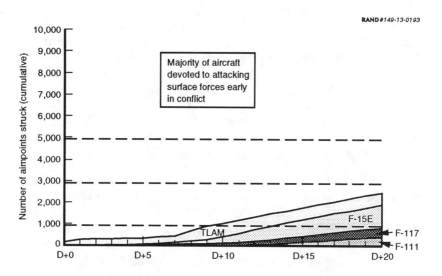

Figure 13—Strategic Attack Potential (Programmed Forces, 1997)

---

[30]The figure does not include the limited number of ALCM-Cs that currently could be delivered by B-52s or other weapons delivered by the B-2 and B-1B.

is established, more of the long-range fighters shift to strategic attacks while shorter-range aircraft (F-16s and F/A-18s) continue to attack enemy forces in the field. The latter aircraft, particularly those equipped for precision attack (e.g., LANTIRN-equipped F-16s), could also contribute to the strategic campaign, but we did not factor these into our calculations.

The F-117s provide a unique and important capability: survivable precision attack of time-critical targets. Early in the campaign, they are the principal contributor to this mission. After the ground situation is stabilized, the majority of the strategic targets are attacked by F-15Es. In our base case (1997), it would take the fighter force approximately eight days to strike the time-critical 250 aimpoints requiring precision weapons and about two weeks to strike the remaining 750 aimpoints requiring precision attack. If the weather more closely resembles the Korean environment (and thus more targets are obscured), it would take about 12 days to strike the 250 time-critical aimpoints and almost three weeks to strike the remaining 750 aimpoints.

TLAMs are very useful weapons. The calculations shown assume that each battle group would salvo its TLAMs over the course of two days; once their magazines were empty, the vessels are assumed to return to a port to reload. Initiatives that increase the number of TLAMs available to the joint force commander would assist in achieving campaign objectives earlier. Clearly, prudent preplanning against critical targets of potential foes—a "conventional SIOP" (Single Integrated Operational Plan)—is needed to ensure that the United States can take full advantage of standoff weapons.

Future Capabilities. The long-range bomber force can make a significant contribution if the United States equips these aircraft with standoff cruise missiles (e.g., a conventional variant of the ALCM, the Tri-Service Standoff Attack Missile [TSSAM], a bomber-launched variant of the TLAM, or other similar weapon) and direct attack weapons, in particular, inertially guided weapons (IGWs). IGWs are simply general purpose bombs (or cluster weapons) equipped with an inertial guidance package. In some ways, these weapons are similar to laser-guided bombs in that the guidance package is strapped on to a general purpose bomb, thus greatly increasing its accuracy for fairly low cost. The bomb's guidance package, which consists of a Global Positioning System (GPS) receiver, an inertial guidance unit, and some steerable fins, is provided with a set of latitudinal and longitudinal coordinates for the target (either before the aircraft takes off or during the mission). When the aircraft is within range of the target (data are provided to the crew by the aircraft's own navigation system and/or GPS satellites, the weapon is dropped and the inertial guidance unit uses aerodynamic fins to steer the weapon to the target's coordinates. IGWs offer an accurate, all-weather day-and-night capability. Laser-guided bombs would still be needed, however, to deal with targets requiring greater precision and/or for a man in the loop.

IGWs and standoff weapons mounted on bombers could greatly increase the rate at which the United States is able to destroy the enemy's war-fighting capabilities. More significantly, the bombers could fill shortfalls in the United States' ability to conduct such attacks from the very outset of a conflict.   In our campaign simulation, the B-2 does not begin strategic attack operations until D+10 because of the need to employ its unique capabilities in attacking maneuver forces (see the following section). B-1Bs provide the dominant punch.

In our simulation, we allocated 75 percent of the 84 PAA B-1B force to this operation (the remaining aircraft provide a reserve to support nuclear deterrence and/or conventional operations in another theater).   The B-1Bs were assumed to initiate operations out of the CONUS and deliver standoff weapons on the first day of combat. After this strike, 60 percent of the allocated B-1Bs were assumed to recover at rear area theater bases (within 3,000 to 4,000 miles of the theater)—the remainder of the force would return to CONUS.[31] From D+0 to D+2, the nonstealthy B-1Bs would continue delivering

---

[31]Conducting heavy bomber operations from bases in the rear of the theater would always be preferred. Theater basing increases sortie rates, reduces tanker requirements, and decreases the complexity of operations. In a short warning scenario, however, launching bombers out of the CONUS for the initial attack and then conducting subsequent operations from theater bases would be the most responsive operational concept—the long range and heavy payload of bombers would provide an important hedge against failures to take action in response to strategic warning. Such missions are certainly possible. During the war in the Gulf, for example, B-52s launched out of the United States to deliver a then classified conventional variant of the Air-Launched Cruise Missile known as the ALCM-C.

The speed at which the United States could establish support for theater operations introduces many important variables and would affect force capabilities. To establish short notice theater operations for heavy bombers, the United States would need to preposition stores of weapons and other equipment at various bases in Europe, Asia, and other locations.

standoff weapons to minimize exposure to enemy defenses. At this point, an increasing percentage of the allocated B-1Bs would begin to penetrate enemy defenses to deliver both gravity bombs and IGWs.[32] After a week of operations, those B-1Bs still operating out of the CONUS would shift their operating locations to rear area theater bases. Table 3 illustrates our B-1B allocation assumptions.[33] Figure 14 illustrates the potential impact of the B-1Bs and B-2s using these assumptions on U.S. capabilities to strike strategic assets. Not only can the United States greatly increase the rate at which it strikes targets, but it can increase its early punch.

Standoff weapons offer high leverage, particularly as more advanced terminal guidance systems and GPS route planning systems currently under development enter service, and allow the United States to strike targets while minimizing the exposure of U.S. personnel to enemy systems.   To deliver such large-scale attacks, the United States clearly needs to improve the joint planning process for target-

Table 3

B-1B Allocation Assumptions (number)

Basing	D+0	D+1	D+2	D+3	D+4	D+5	D+6	D+7	D+8–20
# of B-1Bs CONUS Based (standoff)	64	26	26	26	26	26	26	26	0
# of B-1Bs Theater Based (standoff)	0	38	38	26	26	26	26	26	20
# of B-1Bs Theater Based (Penetrate; IGW)	0	0	0	0	0	0	0	0	44
# of B-1Bs Theater Based (Penetrate; Mk 82)	0	0	0	12	12	12	12	12	0

[32]The magnitude of defense suppression operations required to ensure B-1B surviv-ability in penetrating missions would depend upon the depth of penetration required, the extent to which the enemy's integrated air defense system had been degraded, and the capabilities of the aircraft's self-protection equipment.

[33]The following assumptions were used in the calculations: The sortie rate for CONUS-based bombers was 0.25; for theater-based bombers, 0.5. The mission-capa-ble rates were assumed to be 0.85. B-2s were armed with 16 2,000 lb. IGWs only with an assumed probability of kill (PK) of 0.7.   B-1Bs were armed with a variety of weapons: 8 standoff weapons (with a PK of 0.7; 24 2,000 lb. IGWs with a PK of 0.7) or 84 Mk 82 gravity bombs (with the assumption that each aircraft so armed could strike 8 separate aimpoints, each with a string of about 10 bombs and a PK of 0.6).

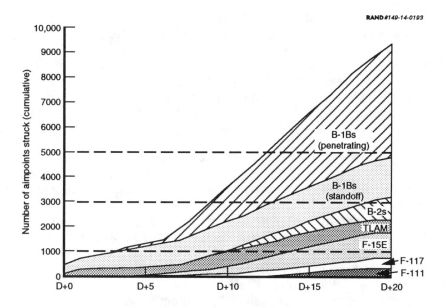

Figure 14—Future Strategic Attack Potential

ing such weapons. IGWs allow the United States to take better advantage of a bomber's weapons bays compared to standoff weapons (since less of the bay's volume is consumed by the fuel and engines needed for standoff weapons). Also, IGWs delivered by penetrating aircraft are far more flexible than standoff weapons, which could prove critical when attacking mobile and dispersed targets, among other things. Finally, fighters delivering precision weapons (with an accuracy of less than ten meters) would be needed to deal with very hard targets (such as command bunkers) and targets where minimizing collateral damage is crucial.

Attacking Enemy Surface Forces

A high priority in many short warning conflicts would be to halt and destroy invading ground forces as quickly as possible. Rapidly stopping the invasion would reduce casualties and lower the cost of retaking territory. In many scenarios, such as the defense of Saudi Arabia or South Korea, stopping the offensive early could mean the difference between preserving or losing critical strategic assets (e.g., Ras Tanura, Dhahran, Seoul).

The capability of airpower to engage and destroy enemy surface forces was demonstrated in the recent war in the Gulf. In the near future, our analysis indicates that U.S. capabilities for destroying enemy vehicles will increase dramatically. The key factors are the introduction of new munitions, avionics, and aircraft; enhanced and rapidly deployable theater surveillance capabilities, such as those provided by JSTARS and other assets; and a rapidly deployable theater C^3I system that can focus firepower where needed.[34]

Figure 15 depicts the overall concept of operations used in our analysis for attacking the invader's forces. Because of the growing capacity to detect enemy forces at great distance and then engage with mass, lethality, and precision, we believe the distinction between "close air support" and "interdiction" is becoming less meaningful. In our concept, air and land forces engage forward, rear, and transiting enemy forces more or less simultaneously. Attacks against enemy lines of communication (such as striking chokepoints and bridges, as well as mining likely avenues of attack) would further hinder an aggressor's offensive. To minimize attrition, the study provided concentrated defense suppression and force protection assets where required.

Figure 16 shows our estimate of the daily kill potential of deployed land-based and sea-based airpower against enemy armored vehicles in our base case scenario. With currently available munitions (CBU-87 cluster bombs and Maverick), the anti-armor capabilities of U.S. aircraft are relatively modest.[35] Arming deployed forces (both USAF and naval assets) with Tactical Munitions Dispensers (TMDs) filled with smart anti-armor submunitions—the Sensor Fuzed Weapon (SFW or CBU-97/B)—dramatically increases airpower's

---

[34]Theater C^3I analysis was conducted off-line to determine the number, type, and capability of required systems. In general, systems currently fielded (with modest upgrades) combined with systems about to become operational (notably JSTARS) would provide the nation with a rapidly deployable theater C^3I system. Only modest investment, much of it already planned, is required to create two such systems for two theaters.

[35]The results depicted are compatible with those achieved in Operation Desert Shield/Storm.

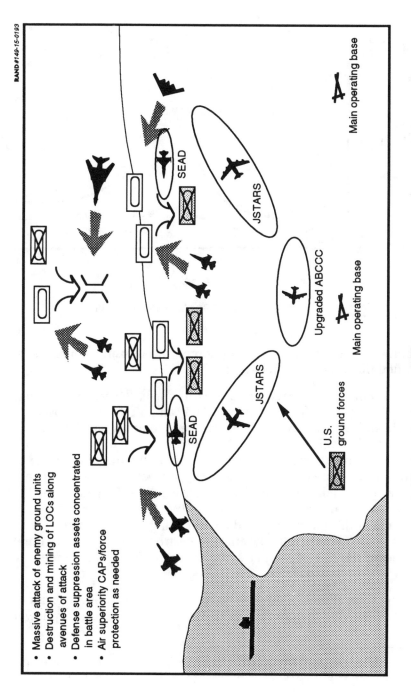

- Massive attack of enemy ground units
- Destruction and mining of LOCs along avenues of attack
- Defense suppression assets concentrated in battle area
- Air superiority CAPs/force protection as needed

RAND#149-15-0193

Main operating base

JSTARS

SEAD

Upgraded ABCCC

Main operating base

U.S. ground forces

JSTARS

SEAD

Figure 15—Attack of Enemy Surface Forces

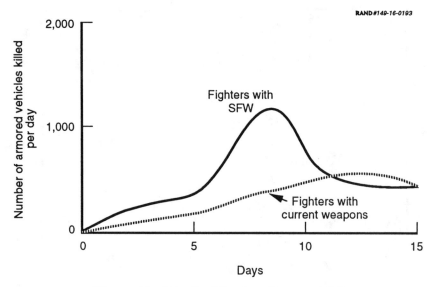

Figure 16—Attack of Surface Forces, 1997

ability to stop moving ground forces, even using conservative assumptions about its effectiveness.[36]

Anti-armor capability reaches a peak in our base case in about ten days. At this point, sufficient attack aircraft have deployed to thetheater to destroy moving armored columns at the rate of hundreds of vehicles per day. Detailed analysis of the ground battle indicates that attacking forces would, after suffering these sorts of attrition, almost certainly be stopped (see below). Once stopped, dispersed, and dug in, armored vehicles would be far less vulnerable to attacks by area munitions, such as SFW. Hence, we switched tactics and at this point employed aircraft using "point" weapons (e.g., laser-guided bombs and Maverick) to destroy vehicles one at a time, reducing the rate of kill per sortie compared to those afforded by SFW against moving forces.

Future Capabilities. Finding ways to stop an armored invasion more quickly remains an important goal. By the turn of the century, it will

---

[36]The effectiveness used for the Sensor Fuzed Weapon is based on test results, detailed ground battle simulations (including countermeasures), and previous RAND studies.

be possible to equip bombers with inertially guided dispensers filled with SFW.[37] Such IG/SFWs mated to the stealthy B-2 would give the United States the capability of attacking invaders with precision and mass virtually from the outset of hostilities—something no other weapons system can offer.

To estimate the capabilities of B-2s armed with IG/SFWs, we conducted a detailed simulation employing these weapons in a highly detailed U.S. Army combat simulation (the JANUS model). In this simulation, we varied target acquisition time consistent with the ability of the B-2 to use JSTARS or on-board sensors to acquire moving armored forces. Additionally, we conducted a sensitivity analysis by varying weapon accuracy, dispenser heading error, submunition search patterns, and enemy force dispersion.[38]

The results shown in Figure 17 are representative of conservative assumptions regarding B-2 effectiveness against enemy-armored for-

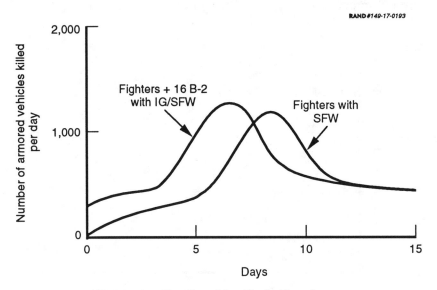

Figure 17—B-2 Provides Early Punch

---

[37]SFW is currently programmed, but the inertially guided TMD is not.

[38]For a detailed description of this analysis, see Glenn Buchan, David Frelinger, and Tom Herbert, *Use of Long-Range Bombers to Counter Armored Invasions*, RAND White Paper, March 1992.

mations. Nevertheless, they show that this concept could provide very substantial anti-armor capabilities at the outset of a conflict, even if only 16 PAA B-2s are fielded. Note that the kill potential peaks earlier than in the base case line illustrating the fighter force contribution—this is because ground battle analysis indicates that the enemy force would have stopped sooner due to the high attrition inflicted by the combination of B-2s and fighters.

ESTABLISHING AN ASSURED DEFENSE

A key initial objective of the joint force commander in the scenario laid out above would be to stop the invasion and establish an "assured" defense. By this we mean inflicting sufficient attrition on the enemy's ground forces so that there is a high probability enemy forces would have to stop their advance and assume a defensive posture. Figure 18 provides a conceptual overview of the process: Enemy ground forces are destroyed by airpower and indigenous ground forces while arriving U.S. ground and air forces add to U.S.

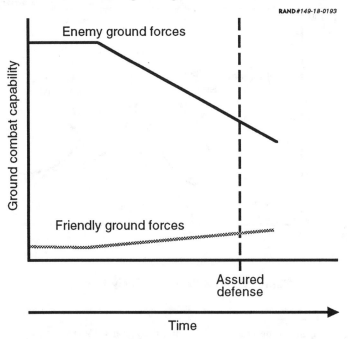

Figure 18—Establishing an Assured Defense

combat power. At some point, an assured defense could be established. We integrated results from our airpower effectiveness model into the RAND-developed evaluation of ground campaigns to determine the time and place when enemy ground forces would be likely to halt. To do this, we used a specific case, an invasion of Kuwait and Saudi Arabia by revitalized Iraqi forces.

This model provides a step-by-step assessment of the outcome of specific ground engagements considering terrain, relative combat power of ground units, movement rates, and the contribution of airpower. The results provided by this model are similar to those provided by other analytic tools, such as the RAND Strategy Assessment System (RSAS). We estimated that an assured defense could be established somewhere between the time when 30 percent of the invading force had been destroyed and when attacking enemy forces had been reduced to equal combat power compared to friendly forces. In the latter more conservative case, about 5,000 armored vehicles out of a total of some 8,500 from the attacking force would need to be destroyed.

We investigated joint force capabilities for a wide range of conditions in both the SWA and Korea scenarios. The SWA scenario was clearly the more challenging due to the severe asymmetry between friendly and enemy ground forces. Figure 19 illustrates when the models indicate that the United States could achieve an assured defense where D-Day equals C-Day. Although these calculations seem to be well calibrated according to U.S. experience in Operation Desert Storm, they should not be regarded as predictive of war outcome. Rather, these represent comparisons of the relative effectiveness of alternative forces in a range of circumstances.

As can be seen, the indigenous ground forces alone in SWA were assessed to be unable to stop the invasion by themselves. Programmed forces, deployed as shown earlier and armed with current munitions, could achieve an assured defense between 9 and 14 days—unfortunately, by this time the enemy could control significant territory and vital facilities. In the event of an invasion of Kuwait and Saudi Arabia, enemy forces might well be able to approach Dhahran before being stopped, which is clearly an unfavorable outcome.

Equipping land-based and sea-based airpower with SFW would permit the United States to establish an assured defense within five to ten days. The range of uncertainty regarding the establishment of

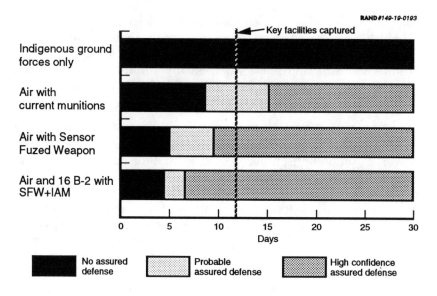

RAND #149-19-0193

NOTE: This chart reflects the contribution of a joint coalition force. The variations in the capability of air power are shown in the labels to the various bars.

Figure 19—Joint Force Capability (D=C)

a defense is significantly reduced. Finally, the bottom bar shows that B-2s, when armed with IG/SFW, provide a significant enhancement to U.S. defense capabilities through massive precision attack early in the war.

Figure 20, drawn from our detailed ground combat simulation, translates the results shown in the previous figure into estimates of Forward Line of Troop (FLOT) locations for each case (using the higher level of attrition required for assured defense). With only in-digenous ground forces to oppose them, enemy forces could enter the oil fields after two weeks and continue forward largely unop-posed. U.S. airpower with current munitions could permit estab-lishment of a defense north of Al Jubail, but critical facilities would be in enemy hands and the margin of safety perilously thin. By em-ploying SFW from land-based and sea-based airpower, critical facili-ties would be protected but Kuwait would be lost. Finally, combining land-based and sea-based fighter forces with the massive firepower of the B-2 would allow the United States to stop enemy forces near the Saudi border while possibly protecting Kuwait City. Again, these

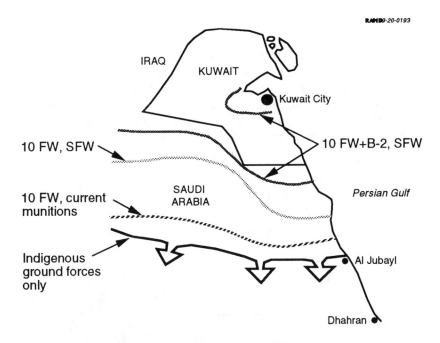

NOTE: This chart reflects variations in the airpower component of a joint coalition force.

Figure 20—FLOT Positions When Assured Defense Achieved (D=C)

simulations should not be taken as predictive of the outcome of future wars. The range of uncertainty as to threat, enemy strategies, U.S. responses, and performance of forces could certainly change the actual outcomes.

These results do not suggest that airpower alone will suffice to provide an assured defense against invading ground forces. Ground forces played an important role. And as shown in some situations in the following analysis, weather, countermeasures, disruptions in the deployment of forces, and enemy operational strategies could work to reduce the effectiveness of an "air dominant" approach. However, it is important to note that the calculus has changed. Lethality improvements available through systems—either here today or about to become operational—have increased the role that air can play in the air-land battle.

Other Excursions

The results displayed previously are only a small set of cases examined in this study. Figure 21 displays the time required to achieve an assured defense for a wider range of cases in the SWA scenario.[39] In these cases, we use the more demanding measure of assured defense: the attrition of enemy ground forces until they are equal in combat capability to defending indigenous ground forces. In Figure 21, we see the effects of the following:

- **Varying the size and effectiveness of USAF power projection forces:** In our base case, ten FWs are deployed. With current munitions, a defense is not attained until almost two weeks have elapsed. It takes even longer when only six FWs are deployed. In

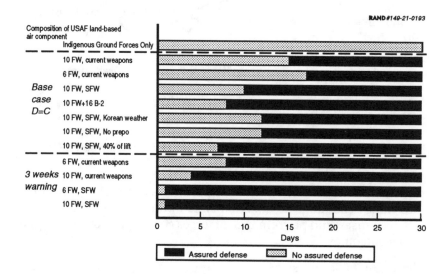

NOTE: Variations in the airpower component are shown in the labels on the left.

Figure 21—Joint Force Capabilities

---

[39]We also ran a case where the enemy armored forces were double the size of that assumed in our base case. In this situation, an air component containing ten FWs and armed with current munitions could not stop the enemy offensive. When armed with CBU-97B, this force could stop the enemy advance in 12 days.

both cases, critical facilities would have been captured. With ten FWs using SFW, the invasion is halted in about ten days, short of the oil fields and ports. Fighters combined with B-2s armed with IG/SFW stop the ground assault even sooner.

- **Weather:** Poorer weather would increase the time required to establish an assured defense. We estimate that if average Korean weather conditions were to prevail in the Gulf region, they might impose a delay of three days or so on the U.S. ability to halt the attack when compared to the base case. In this situation, critical facilities could be lost.

- **Prepositioning constraints:** If prepositioned munitions (both land- and sea-based) were unavailable, more airlift would be needed to move weapons for land-based air forces and/or to accelerate efforts to move in ground forces. This would significantly slow the process of stopping the invasion.

- **Airlift allocation:** Increasing the proportion of airlift dedicated to fighter deployment could reduce the time required to stop an enemy advance. Providing a larger share of airlift to USAF power projection forces in the early phases of the campaign produced a very significant improvement in force performance.

- **Additional carrier forces:** Carriers provide valuable forward presence. They also have the potential advantage of being on the scene early in response to strategic warning and, at least in the initial stages of operations, do not have to rely on access to theater bases. We examined a range of cases in which carrier fighters were the only attack assets employed—for analytic purposes, we assumed the availability of the USAF's C^3I system to focus carrier firepower most effectively and the use of SFW to maximize kill rates. In the first case (with three carriers arriving on C+0, C+7, and C+28, respectively), it would take over a month to establish an assured defense. We also examined two alternatives: one in which four carriers were available and arrived at weekly intervals (C+0, C+7, C+14, and C+21); and one in which four carriers were on station at the start of conflict. In the first alternative, an assured defense could be established in just under four weeks; in the second alternative, the four carriers could establish an assured defense in just over two weeks. While carriers can provide valuable forward presence, naval forces, like ground and air forces, cannot be expected to win wars in isolation.

- **Increased warning times:** To examine the effects of increasing warning time, we assumed a case where three weeks separated C-Day from D-Day. This would allow the United States to deploy more ground and naval forces as well as the entire fighter force before the war begins. We evaluated the time necessary to establish an assured defense under these conditions using previously described forces, but varied USAF fighter force levels. With an Air Force air component of six FWs and current munitions, the RAND model indicates a defense could be established after eight days. When considering the availability of SFW and a USAF air component of either six or ten FWs, a defense could be established in a few days. These results dramatically illustrate that it would clearly be in the interests of potential adversaries to deny the United States useful strategic warning.

ASSESSING AIRPOWER'S EFFECTIVENESS IN THE JOINT CAMPAIGN

Calculating the speed at which the United States can achieve a range of theater objectives provides useful insights into joint force effectiveness. To accomplish this, we combined several measures into one chart to illustrate the capabilities of U.S. forces under differing conditions of accomplishing theater objectives.

In the following evaluations, we measured the time required to (1) strike the 1,000 aimpoints that require precision and (2) concurrently achieve U.S. objectives against enemy surface forces. Figure 22 introduces the format of the figures that follow it. The left-most point of the line on the chart marks the times at which two important objectives are reached:

- The destruction of 250 time-critical aimpoints requiring precision attack; and

- The establishment of an assured defense (i.e., the enemy offensive is halted).

Once these objectives were accomplished, we calculated the time required for the force to reach a further set of objectives:

- The destruction of an additional 750 strategic aimpoints requiring precision attack; and

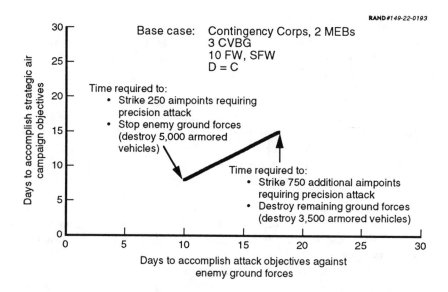

Figure 22—Evaluating Force Effectiveness

- The destruction of the enemy's remaining armored vehicles (which number about 3,500).[40]

The time at which this second pair of objectives is achieved is plotted as the right-most point on the line. In this format, more effective forces are characterized by shorter lines with their left-most points closer to the origin of the two axes.

Figure 22 illustrates the force capabilities in our base case (i.e., SWA, C-Day = D-Day, five divisions, two Marine brigades, three carrier battle groups, and ten FWs with SFW) for accomplishing both sets of objectives.[41] Ten days are required to establish an assured defense, 8 days to strike the 250 time-critical aimpoints. Likewise, 18 days are needed to destroy the rest of the enemy's surface forces; 15 days for

---

[40]To destroy remaining armored vehicles, we calculated the time it would take to destroy each vehicle twice (due to anticipated difficulties in acquiring scattered static land force targets and assessing results) with a combination of Maverick and laser-guided bombs.

[41]During the time required to conduct precision attacks with F-117s, F-15Es, and F-111s, TLAMs and bombers armed with standoff weapons and IGWs would be engaging the other aimpoints requiring less precision.

the precision attack of the remaining 750 aimpoints. Therefore, 18 days are needed to meet the campaign's objectives. It is important to note that the fighter forces in all of these cases also conduct air superiority and SEAD operations.

This format can be used to quickly compare the capabilities of different combinations of force elements, munitions, support assets, and mobility forces. The time required to reach the first set of objectives is a product of the deployability, lethality, and responsiveness of the force. The follow-on objectives are a function of the size of the force used, its lethality, and force allocation strategies.

As can be seen in Figure 23, varying the size of the land-based air component principally affects the ability to meet the final set of objectives because an assured defense can typically be reached before the entire fighter force is deployed (whether six or ten FWs are considered). *When the forces are armed with SFW, the point at which an assured defense can be reached is determined primarily by the rate at which air-to-surface forces, along with the necessary share of force protection and support assets, arrive in the theater.* The time needed to reach the subsequent set of objectives is a function of overall force size. Large numbers of aircraft are needed to search out and destroy enemy ground forces in a rapid manner once a defense is estab-

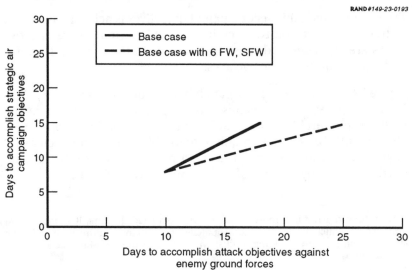

RAND#149-23-0193

Figure 23—Varying the Land-Based Air Component

lished. In each case, 1,000 strategic aimpoints requiring precision delivery are successfully attacked before all objectives relating to enemy surface forces are reached.

A force of ten fighter wings provides a decisive force permitting parallel warfare in the three missions considered here, with the flexibility to increase the effort in any mission where the United States is challenged while maintaining efforts across the board. Put simply, a larger force increases U.S. flexibility in the face of uncertainty. If weapons do not prove as effective as anticipated, more sorties will be needed. If surface-to-air defenses prove resilient, more assets would be available in a larger force to conduct defense suppression. Similarly, should it prove necessary to undertake tasks not accounted for in this analysis, such as "hunting" for mobile ballistic missiles or other critical targets, a larger force would permit this without having to stop prosecuting other aspects of the campaign. Finally, a larger force can reduce the vulnerability of the overall force to disruptions. For example, a concerted enemy attempt to attack fighter bases would probably have less effect on a larger force (which could be spread over more bases) than a smaller force concentrated on fewer bases.

The time required to establish an assured defense is determined primarily by the rate at which forces arrive in the theater. One approach to improving performance in the critical early days would be to increase the amount of airlift dedicated to the deployment of land-based airpower. In the case displayed in Figure 24, an additional 15 percent of the airlift (~12 airlift sorties arriving in theater per day) was tasked to move support equipment, associated C^3I assets, and munitions for fighter units to the theater. This improved the time when a successful defense could be established by three days (though it delayed the closure of the first brigade of the 82nd Airborne Division by several days). There is a limit to how much benefit can be gained through altering the airlift allocation, since fighter force deployment rates are also constrained by tanker availability and aerial refueling capacity.[42]

---

[42]For a detailed analysis of tanker requirements for supporting fighter deployments, see Michael Bednarek, *Alternative Concepts for Aerial Refueling of Deploying Tactical Fighters*, RAND, N-2960-AF, 1990.

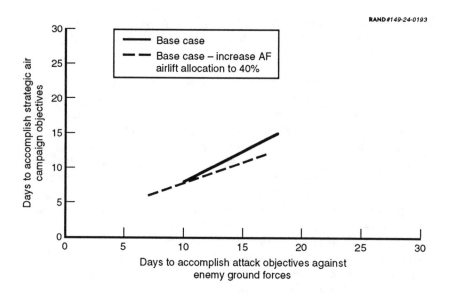

Figure 24—Adjusting Airlift Allocation

As noted earlier, we also examined a situation in which airlift would be constrained in the early days of the deployment at the "ramp up" rate witnessed in Operation Desert Shield. If this constraint were encountered, three additional days would be needed to establish a defense. In the SWA scenario, this delay could mean the loss of critical facilities. But fairly minor adjustments to airlift allocation over the first week emphasizing deployment of land-based airpower assets, would allow the United States to stop the enemy forces as quickly as in our base case.

A far more serious obstacle to meeting the first objective would be the unavailability of land-based or maritime-based prepositioned munitions. In this situation, more airlift would be consumed in carrying munitions. The time required to establish an assured defense would only increase by two days, but because more airlift would be consumed in transporting heavy bomb bodies to the theater, substantially more time would be required to destroy the remaining surface forces and conduct precision strikes of 1,000 strategic aimpoints. This reinforces the crucial importance of prepositioning munitions (maritime prepositioning offers the most flexibility) and establishing a quick-response munitions distribution system for deployment to the area of operations.

The SWA scenario occurs in a region where the weather is generally favorable for the conduct of intense air operations. We employed historical weather patterns in our analysis, but we felt it was important to understand the effects of degraded conditions on the campaign results. In the situation displayed in Figure 25, we estimate the effects of average Korean weather on the ability to achieve operational objectives in SWA, the most stressful case. The effects are predictable. Additional days are needed to achieve our objectives, particularly those dependent upon the employment of laser-guided weapons, which are more vulnerable to poor weather. A similar result might be encountered if an adversary were to employ highly effective countermeasures (camouflage and deception) against the United States. These results illustrate that the possible constraints on the application of airpower (such as poor weather, highly capable enemy air defenses, and munitions availability) dictate against reliance upon an "air only" approach. A joint approach that allows the United States to take full advantage of the unique and interrelated capabilities of all the Services offers the critical flexibility needed to cope with the uncertain future.

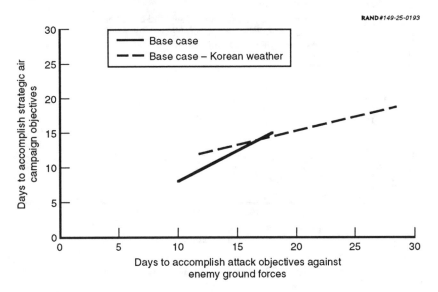

Figure 25—The Effects of Weather

Finally, it is useful to compare the results achieved in our base case to a situation where more warning time is available. The results shown in Figure 26 assume three weeks of warning time and a USAF component of either six or ten fighter wings. In such a case, more ground forces, two carriers, and all fighter wings are available at the start of conflict. As a result, this force can successfully meet the demands of the situation. However, employing a larger fighter force armed with effective munitions decreases the length of time required to meet U.S. objectives.

In analyzing the contributions of various components of the power projection forces, we observed:

• SFW provides very significant increases in anti-armor capability over current munitions against moving ground forces.

• Carrier-based airpower today would probably compose the leading edge of the U.S. fighter attack force at the outset of a conflict but, with current munitions, possesses limited capability to stop attacking surface forces. Arming carrier aircraft with SFW

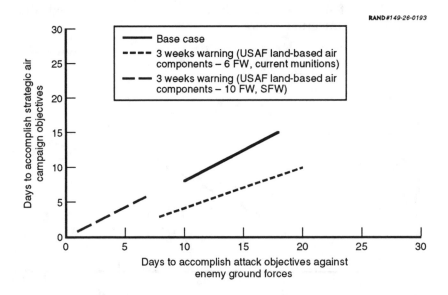

Figure 26—Increased Warning Time

or an equivalent type of weapon would allow them to contribute more effectively to joint operations. In all our calculations with SFW, we assumed that the F/A-18s available were armed with these munitions.

- B-2s launching out of CONUS and then conducting sustained operations from bases near the theater of operations could provide critical firepower in the early days of the campaign.

Finally, the F-15E, with its heavy payload and long range, proved to be the workhorse of the land-based fighter force in our simulations. As we conducted the analysis, it became increasingly apparent that F-15Es played a major role in the destruction of enemy ground forces and, later, the destruction of enemy strategic targets. Figure 27 illustrates the F-15E's potential contribution in attacking either armored vehicles or strategic targets if allocated entirely to these missions (except for 25 percent of the F-15Es withheld for SEAD missions). The time required to execute attacks on surface forces or strategic targets depended heavily on the allocation of the F-15E force. In our base case, for example, where we first emphasized attacking surface vehicles and then emphasized strategic attacks, the F-15Es over the first ten days not only contributed heavily to the SEAD mission, but accounted for more than 40 percent of the armored vehicle kills and more than two-thirds of the precision attack capability for the strategic air offensive campaign.

F-15Es offer a high degree of effectiveness and flexibility to the power projection forces because of their long range, large payload, and modern avionics systems. As we have seen, these aircraft contribute significantly to both the attack of enemy surface forces and the strategic air offensive campaign. This platform's potential capabilities to become a defense suppression asset are also drawn upon in our simulations. Further, the F-15E's sophisticated avionics suite (including a synthetic aperture radar for location and identification of surface targets) could be useful with external cueing to allow the aircraft to conduct special strike/reconnaissance missions for the attack of fleeting mobile targets such as ballistic missiles. The inherent qualities of this aircraft make it a valuable, but numerically limited, asset for an uncertain world.

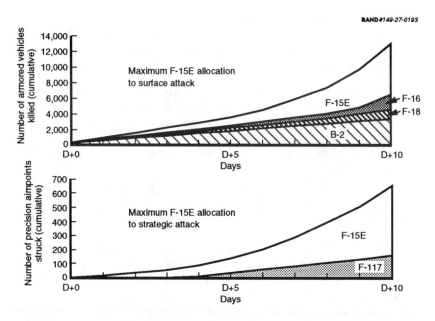

RAND#149-27-0193

NOTE: 25 percent of F-15Es were also allocated to SEAD mission.

Figure 27—F-15E Is Major Contributor

LAUNCHING A GROUND OFFENSIVE

One of our analytical objectives was to assess the capabilities of U.S. forces in operations beyond the defensive phase and to determine when enemy forces had been weakened to the point that allied ground forces could go on the offensive. But when we considered the emerging capability of the combined force, the key constraint appeared to be not the ability to weaken enemy ground forces, but the rate at which U.S. and allied ground forces could build up combat and support forces in the theater.

Figure 28 illustrates the buildup of a fully supported ground force (both Army and Marine) in the theater of operations and is based upon favorable planning assumptions. Light ground forces (the 82nd Airborne and two Marine brigades) would deploy using about half of all available airlift assets (the Marines would also employ maritime lift assets for heavy equipment). They would close in the battle

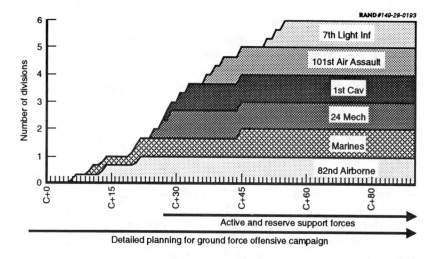

SOURCE: Secretary of Defense Dick Cheney, *Annual Report to the President and Congress*, USGPO, February 1992, pp. 92–99.

Figure 28—It Takes Time to Build Up for Offensive Ground Operations

area as rapidly as is consistent with marshaling capability at departure bases and their ability to organize units and equipment after arrival in the theater. Heavy ground forces would move to CONUS ports before the outbreak of fighting and, using sealift, could begin to arrive in the theater about three weeks after C-Day. A force of this size would require from 180,000 to 200,000 combat and combat service support personnel, 40,000 trucks, and sustainment to conduct offensive ground combat operations in an "immature" theater. Additionally, conducting detailed planning for combat and logistics support of maneuver operations of this scale requires time. We estimate that at least 60 days after C-Day would be required before these forces would be ready to begin a ground offensive under the best of conditions.

The primary focus of our work was to examine Air Force capabilities within a joint force context. As a result, we have not analyzed the wide range of potential improvements available to increase the lethality and mobility of U.S. ground forces (such as lighter armored vehicles, more effective man-portable anti-armor weapons, advanced attack helicopters, etc.). Nonetheless, it is clear that prepositioning of more ground force equipment (both maritime, as with the Marines, and land-based) would reduce the time that heavy U.S. ground forces would need before they would be ready to fight. As highlighted by the Joint Staff's recent mobility requirements analysis, prepositioning clearly should be pursued to bring heavy ground combat power to bear sooner.

ASSESSING CAPABILITIES FOR A SECOND CONCURRENT MRC

We now turn to assessing U.S. capabilities to conduct a second concurrent major regional conflict. For analytic purposes, we assumed the second conflict would be of the same size as the SWA scenario evaluated above. We also varied the times between the outbreak of the first conflict and the onset of the second.

For the first conflict, we committed sufficient forces to ensure a decisive outcome. The Army committed 5 divisions and their required combat support and combat service support, the Navy 3 carrier battle groups and 2 Marine Expedition Brigades (MEBs), and the Air Force 10 fighter wings (including two reserve FWs), 58 heavy bombers, tankers, tactical airlifters, and an array of command and control assets.

Many uncertainties surround the availability of forces for a second conflict. Figure 29 provides a simplified picture of forces from the Base Force that might be committed to an MRC and those that would, in principle, be available for operations elsewhere. However, several factors might restrict or delay the availability of these residual forces. For example, perhaps four fighter wings and three divisions would be stationed overseas, which might place restrictions on their availability for tasking (though some might be located in the right place).

Though a number of Army divisions remain, reserve Army divisions are limited in terms of availability by typical train-up times. Bringing a large, complex unit like a reserve division up to fighting standards can take months. Large formations capable of engaging in maneuver warfare also require a considerable amount of support. As currently structured, the Army would have difficulty generating sufficient combat and combat service support units to conduct operations in

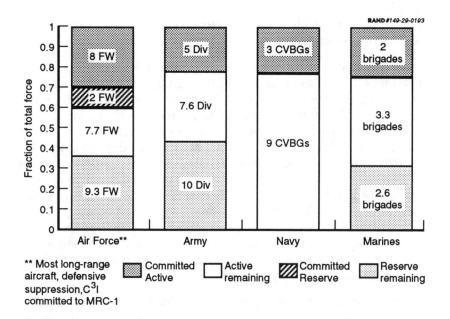

RAND #149-29-0193

** Most long-range
aircraft, defensive
suppression,C^3I
committed to MRC-1

Committed
Active

Active
remaining

Committed
Reserve

Reserve
remaining

Figure 29—Significant Forces Are Available for Second MRC

two theaters in a timely manner for other than light force operations. A commitment of five divisions to two separate theaters, after all, would require generating two separate support elements composed in total of nearly a half million personnel. Cuts below the Base Force level would further restrict availability of forces for concurrent operations.

For short notice conflicts, the three carriers committed above are typically all that might be readily available. Official USN posture documents state that three carriers are required for each carrier forward deployed on a sustained basis—the specific ratio depends on distance from home ports.[43] While the Navy could certainly "surge" a higher proportion of its carriers forward in wartime, such a surge could take months to generate. Under the Base Force, the Marines would have committed 25 percent of their total force to MRC 1 (2 of 8

---

[43]See, for example, *Seapower for a Superpower,* Department of the Navy, Washington, D.C., 1992.

brigades). This would leave 3.3 active brigades available for MRC 2 and 2.6 reserve brigades.

Assuming that forward-deployed wings must be maintained in position for political reasons and that a typical number of wings would be converting to new systems (and hence be unavailable for deployment), the Air Force under the Base Force would retain 10 to 12 wings that might be employed for a second operation.[44] Though many of the wings would be in the Air Reserve Component (ARC) (both National Guard and Reserve), USAF reserve component power projection forces are typically highly prepared and could be ready in less than a month. But as currently constructed, the Air Force would have difficulty fielding a balanced force for a second contingency because of insufficient long-range attack aircraft (such as the F-15E) and inadequate numbers of deployable theater surveillance and command and control systems.[45] It is not a question of force size, but force mix.

Another consideration in selecting forces for a second conflict is the capability of moving them rapidly to the theater using the fastest mobility tool: airlift. Figure 30 and Table 4 illustrate the relative weights of selected Air Force power projection forces (with munitions) and light Army forces. It appears unfeasible to attempt to move a heavy Army division by air. For example, deploying the 3rd Armored Division would require that the United States lift approximately 150,000 tons (not counting combat service support assets)— almost seven times the weight of six fighter wings with nine days of munitions. Therefore, we deemed it impracticable to move heavy Army divisions to a regional conflict by any means other than sealift.

With these factors in mind, we examined a situation in which U.S. forces might be required to deploy to and fight in two concurrent major regional conflicts. The outbreak of hostilities between the first and second conflict were separated, in this case, by 21 days. We also

---

[44]Actual numbers available would depend on logistics support (such as mobility kits and spares). Typically, Air Force wings maintain two independent squadrons (capable of deploying anywhere) and a dependent squadron (which must be paired with an independent squadron when deploying).

[45]We conducted an off-line assessment of C^3I capabilities required to conduct theater operations. Our findings indicate that modest investment in theater C^3I could overcome the latter deficiency.

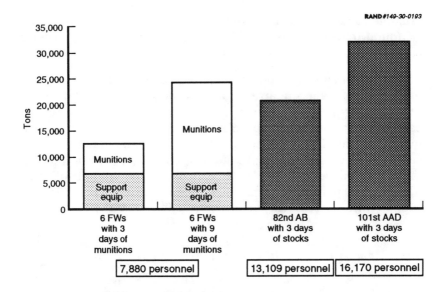

SOURCE: Kenneth Kelley, *Deployment Planning Guide,* Military Traffic Management Command, MTMCTEA Report OA-90-4f-22, August 1991. Estimates for fighter deployment requirements were derived from Unit Type Codes (UTC) mobility packages and average sortie rates and munitions loads as shown in the accompanying table.

Figure 30—Weights of Deploying Forces

Table 4

USAF Squadron Deployment Planning Factors[a]

Type of aircraft	Number of squadrons (24 PAA)	Weight of squadron mobility kits (tons)	Tonnage of munitions per day per aircraft	Number of personnel per squadron
F-15C	5	398	3	422
F-15E	2	378	8	539
F-16	9	314	4	494
EF-111	0.5	347	0	441
F-117	1.5	343	4	356

[a]The factors in this table were used in the calculations for Figure 30. The weight of mobility kits for the various squadrons is an average for independent and dependent squadrons (independent squadrons are slightly heavier than dependent squadrons). Though sortie rates differ depending on aircraft type and length of missions, for purposes of these calculations, fighters were assumed to fly on average 2 sorties per day.

examined a case where only five days separated the two conflicts, but we found that constraints on lift and tankers would make such operations implausible.

The force elements we selected are illustrated in Figure 31. In selecting forces for these concurrent conflicts, we sought to sustain an acceptable war-fighting capability in the first, while maximizing early combat power in the second. We did so by sustaining theater forces in MRC 1 using sealift (which began arriving around day 21), 80 percent of total CRAF (stages I, II, and III), and 20 percent of the committed organic airlift force.[46] This allocation left 20 percent of the CRAF and 80 percent of committed organic airlift assets available to support MRC 2. Critical forces, specifically B-2s, the F-117s,[47] re-

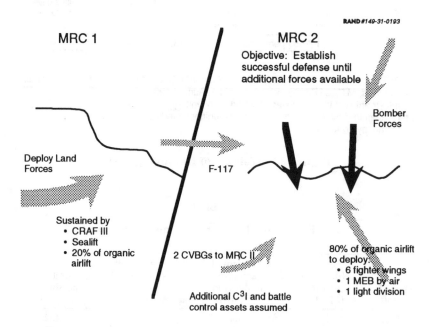

Figure 31—Fighting a Concurrent MRC (21-Day Separation)

---

[46]Ten percent of the total airlift fleet was assumed to be withheld for national missions. The percentages discussed here refer to the proportion of the remaining airlift fleet.

[47]Based on discussions with the Air Staff, we assumed five days would be required to shift the F-117s from MRC 1 to MRC 2.

maining F-15Es not committed to the first MRC,[48] and two carriers previously en route to the first contingency, were shifted from the first to the second conflict.

This is an economy-of-force operation. Six fighter wings were earmarked for the second conflict with sufficient balance to achieve air superiority and to blunt invading ground forces early. Preferred munitions were airlifted into the theater until maritime prepositioning ships arrived (nine days after the conflict erupted). In addition, the simulation airlifted theater missile defense batteries, a light Army division, and personnel from the Marine brigade. To move these forces, we allocated half of the airlift devoted to MRC 2 to the Air Force and the remaining airlift to the other forces. Deployable surveillance and command and control assets for a second conflict, while not currently programmed, were assumed available and factored into the deployment analysis.

While this deployment plan appears feasible in many respects, there are important limitations. These include the availability of tankers, airlifters, and shipping for concurrent deployment and employment in two theaters; the ability of staffs and planning tools to conduct multiple deployment operations; and the ability of the logistics infrastructure to support concurrent operations. In evaluating U.S. capabilities of dealing with two contingencies with D-Days separated by less than three weeks, our analysis indicates that the strains on the tanker and airlift forces alone would prevent the United States from deploying forces to the second conflict in a timely manner.

In the early days of an MRC 2 separated from MRC 1 by three weeks, land-based airpower closes slightly faster than in MRC 1 because somewhat more lift is allocated than in our base case for MRC 1. TMD batteries are deployed at nearly the same rate. The analysis also employed airlift in sufficient numbers to close mobile command and control assets rapidly (AWACS, JSTARS, RC-135s, U-2s, various ground stations, and upgraded EC-130 command and control aircraft). Though the nation does not currently possess sufficient theater command and control assets to prosecute two simultaneous

---

[48]In this situation, the current force composition limited the capability of the force. We also examined a case in which additional F-15Es were available—these aircraft allowed us to achieve our campaign objectives more quickly than in the case shown here.

conflicts, our off-line analysis indicates that modest investment over the next decade would provide the necessary assets.

In the initial stages of combat, almost the same number of dedicated air-to-air assets (five squadrons of F-15Cs and F-14s from two battle groups) are available in MRC 2 as in MRC 1 to establish an air defense. In regard to combat effectiveness in striking strategic targets and dealing with enemy surface forces, Figure 32 illustrates the results of our analysis. These suggest that the United States has the ability to blunt an invasion successfully and conduct strategic strikes in a second conflict. Assets critical to a successful defense include the B-2 and F-15E, advanced anti-armor weapons such as SFW, and a deployable theater C^{3}I system.

Figure 32 indicates the relative capabilities of the force in MRC 2 when compared with MRC 1. Forces allocated to the second conflict are smaller, which means U.S. capabilities for conducting an attack of surface forces and strategic targets simultaneously are reduced. Having held in this theater, the time required to build up additional ground and air forces to eject enemy forces from friendly territory

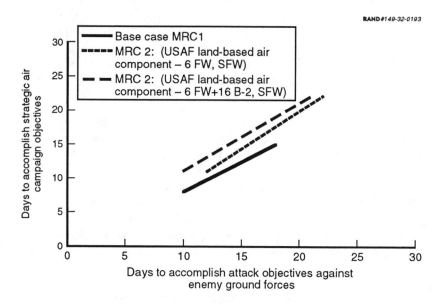

Figure 32—MRC 2 Capabilities

would depend upon the outcome of operations in MRC 1 and the availability of sealift assets to close necessary ground forces to the second conflict.

The force employed in MRC 2 used all the remaining F-15Es in the USAF. A significantly more effective force composition would include four to six squadrons of F-15Es (1.3 to 2 FWs), but since the procurement of these aircraft has been truncated at six squadrons total, the United States could not deploy additional numbers of these critical assets unless some were pulled from MRC 1. Considering the flexibility and combat utility of these aircraft, procurement of additional F-15Es would be a cost-effective means of enhancing USAF capabilities to support national military objectives.

Chapter Four

CONCLUSIONS

The United States will continue to require military capabilities sufficient to deter aggression and to defeat such aggression should deterrence fail. For force planning purposes, it should be assumed that
hostile forces confronted in such conflicts comprise approximately
3,000 to 5,000 tanks and between 500 and 1,000 combat aircraft.
Though challenging, the size of the military threat has decreased
dramatically when compared to the days of the Cold War.

The Joint Chiefs of Staff have recommended that the United States
should field forces capable of defeating aggression in two concurrent
conflicts. Whether or not one believes the probability is very high of
the United States prosecuting two concurrent major regional conflicts, sizing forces for more modest criteria (e.g., for one major regional conflict or for smaller scale conflicts) could engender substantial and unnecessary risks. A larger force structure provides
flexibility and some margin for responding to the unexpected—both
valuable qualities when dealing with something as inherently uncertain as military operations 10 to 20 years into the future.

In addition to the gross *quantitative* criterion of being able to prevail
in two concurrent major regional conflicts, important *qualitative*
criteria should be specified for future U.S. military forces. The
United States must not only be able to achieve its aims in regional
conflicts, but must be able, with high confidence, to win quickly, decisively, and with the capability to minimize casualties. Further, it
must be assumed that economic and manpower constraints, as well
as political sensitivities, will constrain the United States from station-

ing sizable forces overseas on a routine basis. Thus, forces must be rapidly deployable to cope with fast-developing conflicts.

Figure 33 illustrates the contributions over time of the various elements of U.S. joint force posture. In the early stages of crisis, naval forces provide an enduring presence. As the United States moves into conflict, the relative (but not absolute) contribution of naval forces declines; rapidly deployable land-based airpower emerges as the dominant element in the crucial initial stages of conflict. Ground forces build up slowly, but are essential for evicting the aggressor from occupied territory.

U.S. forces in the Base Force have the capability for decisively achieving objectives in a single MRC. Provided sufficient time separates the outbreak of conflict in a second conflict (about three weeks), it appears that the Base Force could also provide the capabilities needed to blunt an invasion in a second conflict.

Other types of military operations, such as insurgencies, coups, counterterrorist actions, or situations requiring naval blockades, are not in any sense "lesser included cases" of major regional conflicts. Hence, full account must be taken of the special requirements im-

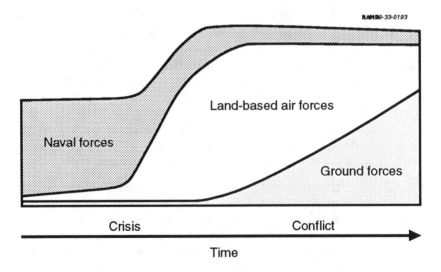

Figure 33—Joint Force Contributions in MRCs

posed by such operations—a task that goes beyond the scope of this study. We also recognize that U.S. forces today lack satisfactory capabilities for destroying small mobile assets, such as ballistic missiles and leadership command posts. As weapons of mass destruction and their delivery means proliferate, this will become an increasingly important problem for U.S. power projection forces. Hence, high-priority should be placed on efforts addressing this shortfall: The United States must not be deterred from intervening to protect important interests abroad.

In posturing its forces to deal with short notice theater conflicts, the United States must rely heavily upon airpower in the crucial initial stages of combat. Aircraft are highly responsive and mobile, capable with tanker and airlift support of deploying anywhere in the world in a matter of days. Such air forces can be supported, at least in the crucial initial stages of combat, by airlift and can outrange almost any opponent through use of the nation's tanker fleet. Though attrition cannot be ignored, judicious employment of electronic and lethal defense suppression systems can minimize losses. Moreover, air operations place at risk a much smaller number of U.S. personnel than do large-scale ground operations.

These results do not imply that airpower alone will suffice to meet the needs of U.S. national security. As illustrated by this analysis, in some situations, weather, terrain, countermeasures, disruptions of the deployment of forces, and enemy operational strategies could work to reduce the effectiveness of an air dominant approach. Other scenarios are certainly possible, and such scenarios would stress different elements of the U.S. joint force structure. An insurgency, for example, would typically demand different sorts of forces: advisory and training missions, civil engineering teams, light ground combat units, helicopter and fixed-wing gunships, and Special Operations Forces. These results imply that the nation needs a joint land, sea, and air force for use in theater conflicts, which together can present potential enemies with the decisive and flexible force needed to underwrite deterrence.

But the results of our analysis do indicate that the calculus has changed and airpower's ability to contribute to the joint battle has increased. Not only can modern airpower arrive quickly where needed, it has become far more lethal in conventional operations. Equipped with advanced munitions either in service or about to become operational and directed by modern C^{3}I systems, airpower has the potential to destroy enemy ground forces either on the move or in defensive positions at a high rate while concurrently destroying vital elements of the enemy's war-fighting infrastructure. In short, the mobility, lethality, and survivability of airpower makes it well suited to the needs of rapidly developing regional conflicts. These factors taken together have changed—and will continue to change—the ways in which Americans think about military power and its application.

To exploit airpower's potential, the United States needs to ensure its ability to control the air, which allows it to conduct more effective attacks of enemy forces and strategic assets. The United States needs to equip its future forces with advanced munitions, which play a critical role in enhancing the lethality of future forces. Stores of these weapons need to be placed on maritime prepositioning ships to increase both flexibility and long-term sustainability. Our analysis indicates that procurement of additional long-range fighters capable of carrying heavy payloads (the F-15E) would significantly increase force effectiveness and flexibility. Finally, a rapidly deployable theater command, control, communications, and intelligence system (consisting of airborne command, control, and surveillance assets combined with deployable ground based facilities) is essential to the effective operations of these forces and appears achievable through the integration of current systems.

Changes in the international environment combined with the increasing effectiveness of U.S. forces mean that reductions in the U.S. military force structure are both possible and prudent. Future U.S. military strategy will set demanding requirements for U.S. military forces. While a smaller force can support U.S. strategy, that force must be of high quality. Hence, the United States must maintain a qualitative edge in its military capabilities through selective modernization. The enhancements discussed above—mobility forces, advanced munitions, advanced fighters, and C^{3}I assets—will require a significant investment. It may be necessary to "trade" a portion of U.S. joint force structure for selective modernization. This will require a new approach to coping with spending cuts, which in the past have focused primarily on reducing procurement accounts and have tended to be apportioned more or less evenly across services and mission areas.